Exosomal Elixir: Unleashing the Healing Potential of Mesenchymal Stem Cell-Derived Exosomes.

ABOUT THE AUTHOR

Meet Dr. James Utley, PhD. He's not your typical Immunohematology guru. For over two decades, he's been crafting a niche in cellular therapy across the U.S. A proud alum of Johns Hopkins, he took a detour to the Department of Defense, where he reimagined the whole cellular transfusion scene.

But here's where it gets intriguing: James is a Biohacker at heart. He's been on the cutting edge, self-experimenting and pushing the boundaries of CRISPR and genetic engineering like a true avant-garde scientist.

Then there's his stint at Banner Health. Not just any role, mind you. He was the Technical Director of all the blood banks and transfusion services, overseeing a monumental 150K successful cellular transfusions with his artisanal touch. But James isn't just about numbers. He's penned thoughts in some of the most avant-garde medical journals and has been the force behind some truly innovative FDA-approved breakthroughs.

And here's the kicker: this former Navy Scientist traded in his lab coat to chart the unexplored waters of the stem cell revolution. Some even call him the "pirate" of the cellular world. Now, as the Chief Scientific Officer at Auragens, he's making waves and a difference in the world every single day. Cheers to the unconventional trailblazer and world-changer!

About the Co-Author

With over twenty years of experience as a realistic visionary and a beneficial entrepreneur, Dr. Dan Briggs is the Founder, President, and CEO of multiple successful companies.

In the medical field Dan is the Founder, President & CEO of Auragens, the premier stem cell and research center in the world and located in Panama City, Panama. Spanning the entire 48th floor of the Oceania Business Tower, Auragens has created the world's leading research facility and has attracted top PhD's and medical doctors from the Americas and Europe in its pursuit to improve the standard of care and treatment. He formed a 501c3 nonprofit, primary care company, The Neighborhood Clinic, where he also serves as Chairman and CEO. With multiple locations The Neighborhood Clinic's doctors, and nurse practitioners, can see up to 500 underserved and rural patients per day that previously had no access to medical care. Dan also founded MDX Labs Inc, a CLIA and COLA Certified, high complexity lab network headquartered in Henderson, Nevada. A frontrunner and innovator in the molecular & clinical diagnostics space, MDX was recognized as "Top Small Business" and "Best Workplace" in Nevada. Additionally, and to further his passion in supporting the community Dan founded and serves as Chairman of FOUNDA-TIONS, a 501c3 charitable organization, that provides support and donations to charities globally assisting those in need and breaking down barriers in access to healthcare. In 2023, Dr. Dan Briggs was named Healthcare Executive of the Year by Nevada Business Magazine and a jury of his peers.

Along with his role in his companies, Dan also has served, and continues to serve, on several boards where he acts as advisor, trustee, and mentor. These include NeoVolta (NASDAQ: NEOV), University of Northwestern Ohio (UNOH), Las Vegas HEALS, and Big Brothers Big Sisters.

Dan's career has firm roots in public health, policy, and advocacy, long-held personal passions of his. Dan was a member of the advance team of President George W. Bush during the first US and Russia Summit with Vladimir Putin, held in St. Petersburg, Russia. Dan then oversaw campaigns on behalf of the Governor of California for the Office of Members Services. After relocating to Nevada, Dan was a Founding Member of the Las Vegas World Affairs Council -- a bipartisan organization dedicated to engaging and educating Americans on international affairs and foreign policy. His desire for public service resulted in him being recruited to run as a candidate for the Nevada State Assembly District 20 in 2008. He lost.

Dan earned his Doctorate Degrees in Public Health (DrPH) and Doctor of Laws (LLD) from University of Northwestern Ohio, studied law at Thomas Jefferson School of Law in San Diego, earned his master's degree in Russian, East European, and Eurasian Studies at Stanford University, and an undergraduate degree in Political Science from Pepperdine University. He is married with three children and splits his time between the USA and Panama.

DISCLAIMER

The content provided in this book is for informational and educational purposes only and is not intended as, nor should it be considered a substitute for, professional medical advice. Do not use the information in this book for diagnosing or treating any medical or health condition. If you have or suspect you have a medical problem, promptly consult your professional healthcare provider. Always seek the advice of your physician or other qualified health provider with any questions you may have regarding a medical condition. Never disregard professional medical advice or delay in seeking it because of something you have read in this book.

CONTENTS

INTRODUCTION TO EXOSOMES AND MESENCHYMAL STEM CELLS

Understanding Exosomes

Exosomes are small, membrane-bound vesicles that are released by various cells, including mesenchymal stem cells (MSCs)[1]. These tiny structures, measuring only about 30-150 nanometers in diameter, play a crucial role in cell-to-cell communication and have gained significant attention in the field of regenerative medicine and therapeutic applications[2].

Exosomes are formed within the endosomal compartment of cells through a process known as inward budding of the endosomal membrane. This results in the formation of multivesicular bodies (MVBs) that contain numerous intraluminal vesicles, which are eventually released as exosomes upon fusion of MVBs with the plasma membrane[3]. Once released, exosomes can travel through bodily fluids, such as blood, urine, and cerebrospinal fluid, to reach target cells and deliver their cargo.

The cargo of exosomes is composed of various bioactive molecules, including proteins, lipids, nucleic acids (such as DNA, mRNA, and microRNAs), and other signaling molecules. These molecules are selectively packaged into exosomes and can vary depending on the cell type and physiological conditions. The specific composition of exosomal cargo determines their functional properties and their potential therapeutic applications[4].

Mesenchymal stem cells (MSCs) are a type of adult stem cell that can be isolated from various tissues, such as bone marrow, adipose tissue, and umbilical cord. MSCs have the ability to differentiate into multiple cell types, including bone, cartilage, and fat cells. In addition to their differentiation potential, MSCs also possess immunomodulatory and regenerative properties, making them an attractive source for exosome production[1].

MSC-derived exosomes have shown great promise in various clinical applications. These tiny vesicles have been found to promote tissue regeneration, modulate immune responses, and enhance the therapeutic effects of MSCs[2]. Exosomes derived from MSCs have been investigated for their potential use in regenerative medicine, neurological disorders, cardiovascular diseases, and immune modulation[4].

In regenerative medicine, MSC-derived exosomes have been shown to promote tissue repair and regeneration by stimulating the proliferation and differentiation of resident stem cells, promoting angiogenesis (the formation of new blood vessels), and reducing inflammation. These properties make them a potential therapeutic option for conditions such as wound healing, tissue damage, and degenerative diseases[1].

In neurological disorders, MSC-derived exosomes have shown promise in promoting neuronal survival, enhancing neuroplasticity, and reducing inflammation. These properties make them a potential therapeutic option for conditions such as stroke, traumatic brain injury, and neurodegenerative diseases like Alzheimer's and Parkinson's[3].

In cardiovascular diseases, MSC-derived exosomes have been found to promote angiogenesis, reduce scar formation, and improve cardiac function. These properties make them a potential therapeutic option for conditions such as myocardial infarction (heart attack) and heart failure[2].

In immune modulation, MSC-derived exosomes have been shown to suppress immune responses and promote immune tolerance. These properties make them a potential therapeutic option for conditions such as autoimmune diseases and organ transplantation[4].

To harvest and manufacture exosomes from MSCs, several methods have been developed. The most common method involves isolating MSCs from the desired tissue source and culturing them in a controlled environment. Once the MSCs reach a certain confluence, the culture media is collected and subjected to a series of centrifugation and filtration steps to remove cellular debris and isolate the exosomes. Additional purification steps, such as ultracentrifugation or density gradient centrifugation, may be employed to obtain highly pure exosome preparations[3].

Manufacturing exosomes for clinical use involves stringent quality control measures to ensure safety and efficacy. The production process must adhere to good manufacturing practices (GMP) guidelines to maintain consistency and minimize batch-to-batch variability. Quality control tests, such as nanoparticle tracking analysis, electron microscopy, and proteomic analysis, are performed to characterize the size, morphology, and cargo of the exosomes. These measures ensure that the final exosome product meets the required standards for clinical applications[2].

In summary, exosomes derived from MSCs hold immense potential for various clinical applications. Their ability to modulate cell-to-cell communication and deliver bioactive molecules makes them a promising tool in regenerative medicine, neurological disorders, cardiovascular diseases, and immune modulation. The harvesting and manufacturing processes of exosomes involve isolating MSCs, culturing them, and purifying the exosomes. Strict quality control measures are implemented to ensure the safety and efficacy of exosome-based therapies[4].

Mesenchymal Stem Cells

Mesenchymal stem cells (MSCs) are a type of adult stem cell that hold immense potential for regenerative medicine and therapeutic applications[5]. These cells can be found in various tissues throughout the body, including bone marrow, adipose tissue, and umbilical cord tissue[6]. MSCs have the remarkable ability to differentiate into multiple cell types, such as bone cells, cartilage cells, and fat cells[7]. In addition to their differentiation capacity, MSCs also possess unique properties that make them an attractive source for the production of exosomes[8].

Exosomes are small extracellular vesicles that are released by cells, including MSCs, and play a crucial role in intercellular communication[5]. These tiny vesicles are packed with a diverse range of bioactive molecules, including proteins, lipids, and nucleic acids, which can be transferred to recipient cells to modulate their behavior and function[6]. Exosomes derived from MSCs have gained significant attention in recent years due to their therapeutic potential in various clinical applications[7].

One of the key advantages of using exosomes derived from MSCs is their ability to mimic the therapeutic effects of the parent stem cells without the risks associated with cell-based therapies[8]. Unlike MSCs, exosomes can be easily isolated, purified, and stored, making them a more practical and scalable option for clinical use[5]. Furthermore, exosomes derived from MSCs have shown promising results in preclinical and clinical studies, demonstrating their potential in regenerative medicine, neurological disorders, cardiovascular diseases, and immune modulation[6].

In regenerative medicine, exosomes derived from MSCs have shown the ability to promote tissue repair and regeneration[7]. These exosomes can stimulate the proliferation and differentiation of resident stem cells at the site of injury, leading to the formation of new functional tissue[8]. Additionally, they can modulate the immune response and reduce inflammation, creating a favorable environment for tissue healing[5]. Clinical trials are underway to explore the use of MSC-derived exosomes in the treatment of conditions such as osteoarthritis, wound healing, and tissue damage caused by ischemia[6].

In the field of neurological disorders, MSC-derived exosomes have shown promise in promoting neuronal survival, enhancing neuroplasticity, and reducing neuroinflammation[7]. These exosomes can deliver neuroprotective factors and promote the regeneration of damaged neurons, offering potential therapeutic options for conditions such as stroke, traumatic brain injury, and neurodegenerative diseases like Alzheimer's and Parkinson's[8].

Cardiovascular diseases, including myocardial infarction and heart failure, represent a significant burden on global health[5]. MSC-derived exosomes have demonstrated the ability to improve cardiac function, promote angiogenesis, and reduce scar formation in preclinical models[6]. These exosomes can deliver growth factors and microRNAs that stimulate the regeneration of damaged cardiac tissue and enhance blood vessel formation[7]. Clinical trials are underway to evaluate the safety and efficacy of MSC-derived exosomes in cardiac repair and regeneration[8].

Another exciting area of research is the use of MSC-derived exosomes for immune modulation[5]. These exosomes possess immunomodulatory properties that can regulate the activity of immune cells and modulate the immune response[6]. They have shown potential in the treatment of autoimmune diseases, graft-versus-host disease, and organ transplantation[7]. MSC-derived exosomes can suppress the activation of immune cells, promote the generation of regulatory T cells, and reduce inflammation, thereby restoring immune homeostasis[8].

The harvesting and manufacturing of MSC-derived exosomes involve several steps to ensure their purity, quality, and therapeutic efficacy[5]. MSCs are first isolated from the desired tissue source, such as bone marrow or adipose tissue, using established protocols[6]. These cells are then cultured under specific conditions to expand their numbers while maintaining their stem cell properties[7]. Once a sufficient number of MSCs are obtained, they are stimulated to release exosomes into the culture medium[8].

The culture medium containing the secreted exosomes is collected and subjected to a series of centrifugation and filtration steps to remove cellular debris and isolate the exosomes[5]. Additional purification techniques, such as ultracentrifugation, density gradient centrifugation, or size exclusion chromatography, may be employed to further enrich the exosome population[6]. The isolated exosomes are then characterized to confirm their identity, size, and cargo composition[7].

To ensure the safety and quality of MSC-derived exosomes for clinical use, rigorous quality control measures are implemented[8]. These include testing for sterility, endotoxin levels, and absence of contaminants[5]. The exosomes are also evaluated for their potency and stability to ensure consistent therapeutic efficacy[6]. Manufacturing processes are conducted under strict guidelines, such as Good Manufacturing Practices (GMP), to maintain product consistency and minimize the risk of contamination[7].

In conclusion, mesenchymal stem cells (MSCs) are a valuable source for the production of exosomes with immense therapeutic potential[8]. These exosomes can be isolated, purified, and stored, making them a practical option for clinical use[5]. MSC-derived exosomes have shown promise in various clinical applications, including regenerative medicine, neurological disorders, cardiovascular diseases, and immune modulation[6]. The harvesting and manufacturing processes of MSC-derived exosomes involve careful isolation, purification, and quality control measures to ensure their safety and efficacy[7]. Continued research and clinical trials are needed to fully explore the therapeutic potential of MSC-derived exosomes and bring them closer to widespread clinical use[8].

Conclusion

Exosomes derived from MSCs hold immense potential for various clinical applications, including regenerative medicine, neurological disorders, cardiovascular diseases, and immune modulation. The isolation and characterization of exosomes are crucial steps in ensuring their quality and efficacy. Techniques such as ultracentrifugation, density gradient centrifugation, size exclusion chromatography, and immunoaffinity capture are employed for exosome isolation. Characterization techniques such as TEM, DLS, Western blotting, flow cytometry, and mass spectrometry provide valuable insights into the size, morphology, and protein composition of exosomes. The manufacturing process involves the cultivation of MSCs under optimized conditions to promote exosome secretion, followed by the isolation of exosomes from the cell culture supernatant. These exosomes can then be further processed and utilized for therapeutic purposes. With ongoing research and advancements in exosome manufacturing, the potential of exosomal therapies continues to expand, offering new hope for the treatment of various diseases and conditions.

THE SIGNIFICANCE OF EXOSOMES IN CLINICAL APPLICATIONS

Exosomes, small extracellular vesicles secreted by various cell types, including mesenchymal stem cells (MSCs), have gained significant attention in the field of regenerative medicine and therapeutic applications. These tiny vesicles, ranging in size from 30 to 150 nanometers, play a crucial role in cell-to-cell communication and have shown immense potential in various clinical applications. Understanding Exosomes Exosomes are nanosized vesicles that are released by cells into the extracellular space. They are formed through the inward budding of the endosomal membrane, resulting in the formation of multivesicular bodies (MVBs). These MVBs can either fuse with the cell's plasma membrane, releasing the exosomes into the extracellular space, or they can be degraded within the cell. Exosomes are enriched with various bioactive molecules, including proteins, lipids, nucleic acids, and signaling molecules. These cargo molecules are selectively packaged into exosomes and play a crucial role in their biological functions. The cargo composition of exosomes can vary depending on the cell type, physiological state, and environmental conditions.[9]

Mesenchymal Stem Cells Mesenchymal stem cells (MSCs) are a type of adult stem cell that can be isolated from various tissues, including bone marrow, adipose tissue, and umbilical cord. MSCs have the remarkable ability to differentiate into multiple cell types, such as bone cells, cartilage cells, and fat cells. In addition to their differentiation potential, MSCs possess immunomodulatory and regenerative properties, making them an attractive candidate for cell-based therapies.[10]

The Significance of Exosomes in Clinical Applications Exosomes derived from MSCs have emerged as a promising alternative to cell-based therapies. These exosomes inherit the therapeutic properties of their parent MSCs and can exert their effects through the transfer of bioactive molecules to recipient cells. The cargo molecules within MSC-derived exosomes can modulate various cellular processes, including cell proliferation, differentiation, immune response, and tissue regeneration.[11]

One of the significant advantages of using exosomes instead of whole cells is their ability to bypass many of the limitations associated with cell-based therapies. Exosomes are non-living entities, which eliminates the risk of uncontrolled cell proliferation or immune rejection. They can also be easily stored, transported, and administered to patients, making them a more practical option for clinical applicatio ns.[12]

Clinical Applications of MSC-Derived Exosomes The therapeutic potential of MSC-derived exosomes has been explored in various clinical applications, including regenerative medicine, neurological disorders, cardiovascular diseases, and immune modulation. In regenerative medicine, MSC-derived exosomes have shown promising results in promoting tissue repair and regeneration. These exosomes can stimulate the proliferation and differentiation of resident stem cells at the site of injury, leading to enhanced tissue healing. Additionally, the cargo molecules within exosomes can modulate the local immune response, reducing inflammation and promoting a favorable environment for tissue regeneration.[9]

In neurological disorders, MSC-derived exosomes have demonstrated neuroprotective and neuroregenerative effects. The cargo molecules within exosomes can promote neuronal survival, enhance synaptic plasticity, and stimulate the growth of new neurons. This has potential implications for the treatment of neurodegenerative diseases, stroke, and spinal cord injuries.[10]

Cardiovascular diseases, such as myocardial infarction and heart failure, have also been targeted using MSC-derived exosomes. These exosomes can promote angiogenesis, reduce scar formation, and improve cardiac function. The cargo molecules within exosomes can activate pro-survival pathways in damaged cardiac cells, leading to tissue repair and functional recovery.[11]

In immune modulation, MSC-derived exosomes have shown immunosuppressive and anti-inflammatory effects. These exosomes can modulate the activity of immune cells, such as T cells and macrophages, leading to the suppression of excessive immune responses. This property of MSC-derived exosomes has potential applications in the treatment of autoimmune diseases and graft-versus-host disease.[12]

Overview of Exosome Harvesting and Manufacturing Processes
The isolation and manufacturing of MSC-derived exosomes involve several steps to ensure their purity, quality, and therapeutic efficacy. The process typically starts with the isolation of MSCs from the desired tissue source, followed by their expansion in culture. Once a

sufficient number of MSCs are obtained, they are stimulated to release exosomes by subjecting them to specific conditions, such as hypoxia or treatment with growth factors.

The conditioned media containing the released exosomes are then collected and subjected to a series of centrifugation and filtration steps to remove cellular debris and larger vesicles. The resulting exosome-enriched fraction is further purified using techniques like ultracentrifugation, density gradient centrifugation, or size exclusion chromatography. These purification steps help to obtain a highly pure population of exosomes.

To ensure the quality and consistency of exosome products, various characterization techniques are employed. These techniques include electron microscopy, nanoparticle tracking analysis, flow cytometry, and proteomic analysis. These methods allow researchers to assess the size, morphology, concentration, and cargo composition of the isolated exosomes.

In conclusion, MSC-derived exosomes hold immense potential in clinical applications due to their ability to modulate cellular processes and promote tissue regeneration. These tiny vesicles offer a promising alternative to cell-based therapies, overcoming many of the limitations associated with whole-cell transplantation. With further research and development, MSC-derived exosomes have the potential to revolutionize the field of regenerative medicine and provide novel therapeutic options for various diseases and conditions.

OVERVIEW OF EXOSOME HARVESTING AND MANUFACTURING PROCESSES

E xosomes, small extracellular vesicles secreted by various cell types including mesenchymal stem cells (MSCs), have gained significant attention in the field of regenerative medicine and therapeutic applications. These tiny vesicles, ranging in size from 30 to 150 nanometers, play a crucial role in intercellular communication and have shown immense potential in various clinical applications. In this section, we will provide an overview of the processes involved in harvesting and manufacturing exosomes derived from MSCs.

Isolation of Exosomes

The first step in obtaining exosomes from MSCs involves their isolation from the cell culture supernatant or other biological fluids. Several methods have been developed for exosome isolation, each with its own advantages and limitations. Ultracentrifugation, one of the most commonly used techniques, involves multiple rounds of centrifugation at high speeds to pellet the exosomes. This method effectively separates exosomes from other cellular debris and contaminants. However, it is time-consuming and may lead to loss of exosomal integrity.

Another widely used method is the use of density gradient centrifugation, where exosomes are separated based on their buoyant density. By layering the sample onto a density gradient medium and subjecting it to ultracentrifugation, exosomes can be isolated at specific density fractions. This technique provides higher purity but is also time-consuming and requires specialized equipment.

In recent years, alternative methods such as size exclusion chromatography (SEC) and immunoaffinity capture have gained popularity. SEC separates exosomes based on their size, allowing for a more rapid and gentle isolation process. Immunoaffinity capture, on the other hand, utilizes specific antibodies to target exosomal surface markers, enabling highly specific isolation. These methods offer advantages in terms of simplicity, speed, and preservation of exosomal integrity.

Characterization of Exosomes

Once isolated, exosomes need to be characterized to confirm their identity and quality. Various techniques are employed to assess the size, morphology, and protein composition of exosomes. Transmission electron microscopy (TEM) is commonly used to visualize the morphology and size of exosomes. This technique allows for direct observation of the vesicles and provides valuable information about their structural integrity.

Dynamic light scattering (DLS) is another widely used method for determining the size distribution of exosomes. By measuring the fluctuations in light scattering caused by the movement of exosomes, DLS provides information about their hydrodynamic diameter. Nanoparticle tracking analysis (NTA) is a similar technique that tracks the Brownian motion of exosomes to estimate their size and concentration.[14]

To further characterize exosomes, techniques such as Western blotting, flow cytometry, and mass spectrometry are employed. Western blotting allows for the detection of specific exosomal markers, such as CD9, CD63, and CD81, confirming their presence in the isolated vesicles. Flow cytometry can be used to analyze the surface markers of exosomes and assess their purity. Mass spectrometry, a powerful technique for protein identification, can provide a comprehensive analysis of the protein content within exosomes.[13]

Manufacturing of Exosomes

The manufacturing process for exosomes derived from MSCs involves the cultivation of these cells under controlled conditions to promote exosome secretion. MSCs can be obtained from various sources, including bone marrow, adipose tissue, and umbilical cord tissue. Once isolated, MSCs are expanded in culture to obtain a sufficient number of cells for exosome production.[14]

To enhance exosome yield and quality, various factors such as culture media composition, oxygen levels, and growth factors are optimized. The culture media are often supplemented with exosome-friendly components, such as exosome-depleted fetal bovine serum, to minimize contamination and maximize exosome production. Additionally, the use of bioreactors and microcarrier systems has been explored to scale up exosome production for clinical applications.

After the MSCs have been cultured and exosomes have been secreted into the supernatant, the isolation process described earlier is employed to harvest the exosomes. Once isolated, the exosomes can be stored at low temperatures or processed further for therapeutic applications.[13]

EXOSOMAL CARGO

Composition of Exosomal Cargo

E xosomes are small extracellular vesicles that are released by various cell types, including mesenchymal stem cells (MSCs)[15]. These tiny vesicles play a crucial role in intercellular communication and have gained significant attention in the field of regenerative medicine[16]. The cargo carried by exosomes is responsible for their therapeutic potential and has been extensively studied to understand their mechanisms of action[17].

Exosomal cargo refers to the diverse array of molecules encapsulated within the exosomes, including proteins, lipids, nucleic acids, and various bioactive molecules. The composition of exosomal cargo is highly dynamic and can vary depending on the cell type, physiological state, and environmental conditions[18]. The cargo is selectively packaged into exosomes through complex sorting mechanisms, ensuring the delivery of specific molecules to target cells[15].

Proteins

Proteins are one of the major components of exosomal cargo and play a crucial role in mediating intercellular communication. Exosomes derived from MSCs are enriched with a variety of proteins that are involved in various biological processes. These proteins include surface markers such as CD9, CD63, and CD81, which are commonly used for exosome identification and isolation[16]. Additionally, MSC-derived exosomes contain proteins involved in cell adhesion, immune modulation, tissue repair, and angiogenesis. These proteins can exert their therapeutic effects by interacting with specific receptors on target cells and activating signaling pathways[17].

Lipids

Lipids are another important component of exosomal cargo. MSC-derived exosomes are enriched with various lipid species, including cholesterol, sphingomyelin, and phospholipids. These lipids not only contribute to the structural integrity of exosomes but also play a role in membrane fusion and cellular uptake. Lipids can also act as signaling molecules and modulate cellular processes in recipient cells[18].

Nucleic Acids

Exosomes derived from MSCs contain different types of nucleic acids, including DNA, mRNA, microRNA (miRNA), and non-coding RNAs. These nucleic acids can be transferred to recipient cells and regulate gene expression, thereby influencing cellular functions. The presence of nucleic acids in exosomes suggests their potential role in genetic information transfer and modulation of cellular processes[15].

Bioactive Molecules

In addition to proteins, lipids, and nucleic acids, exosomes derived from MSCs carry various bioactive molecules, such as growth factors, cytokines, and chemokines. These molecules are involved in cell signaling and can regulate cellular processes, including cell proliferation, differentiation, and immune modulation. The presence of bioactive molecules in exosomes suggests their potential therapeutic applications in tissue regeneration and disease modulation[16].

The composition of exosomal cargo is influenced by various factors, including the physiological state of the MSCs, the microenvironment, and the specific stimuli applied during exosome production. Understanding the composition of exosomal cargo is crucial for harnessing their therapeutic potential and developing effective clinical applications[17].

Role of Exosomal Cargo in Cell Communication and Tissue Regeneration

Exosomes, small extracellular vesicles secreted by various cell types including mesenchymal stem cells (MSCs), play a crucial role in cell communication and tissue regeneration[18]. These tiny vesicles are packed with a diverse cargo of proteins, lipids, nucleic acids, and other bioactive molecules that enable them to transfer information and exert therapeutic effects on target cells. Understanding the role of exosomal cargo in cell communication and tissue regeneration is essential for harnessing the full potential of MSC-derived exosomes in clinical applications[15].

Exosomal Cargo: A Multifaceted Communicator

The cargo carried by exosomes is responsible for their diverse functions in intercellular communication. The composition of exosomal cargo can vary depending on the cell type, physiological state, and environmental cues. The cargo includes proteins, such as growth factors, cytokines, enzymes, and signaling molecules, which can modulate cellular processes and promote tissue repair and regeneration. Additionally, exosomes contain various types of nucleic acids, including messenger RNA (mRNA), microRNA (miRNA), and other non-coding RNAs, which can be transferred to recipient cells and regulate gene expression[16].

Cell Communication and Tissue Regeneration

Exosomes derived from MSCs have been shown to play a crucial role in cell communication and tissue regeneration. These vesicles act as messengers, shuttling bioactive molecules between cells and influencing their behavior. By transferring their cargo to target cells, MSC-derived exosomes can modulate cellular processes such as proliferation, differentiation, migration, and immune response. This intercellular communication mediated by exosomes is vital for tissue regeneration and repair[17].

One of the key mechanisms by which MSC-derived exosomes promote tissue regeneration is through their ability to stimulate the proliferation and differentiation of resident stem cells. The cargo carried by exosomes can activate signaling pathways in recipient cells, leading to the activation of regenerative processes. For example, exosomal cargo can promote the differentiation of stem cells into specific cell lineages, such as osteoblasts for bone regeneration or cardiomyocytes for cardiac repair[18].

Furthermore, MSC-derived exosomes have been shown to possess anti-inflammatory and immunomodulatory properties. The cargo carried by these exosomes can regulate the immune response by suppressing the activation of immune cells and promoting the generation of regulatory immune cells. This immunomodulatory effect is particularly relevant in conditions where inflammation plays a detrimental role in tissue damage, such as autoimmune diseases or organ transplantation[15].

Therapeutic Applications of Exosomal Cargo

The unique properties of exosomal cargo make MSC-derived exosomes promising candidates for various therapeutic applications. The ability of exosomes to transfer bioactive molecules to target cells opens up possibilities for targeted drug delivery and gene therapy. By loading exosomes with specific cargo, it is possible to enhance their therapeutic potential and direct their effects towards specific cell types or tissues[16].

In the field of regenerative medicine, MSC-derived exosomes hold great promise. These vesicles can be used to promote tissue repair and regeneration in various conditions, including bone and cartilage defects, cardiovascular diseases, neurodegenerative disorders, and wound healing. The cargo carried by exosomes can stimulate the regeneration of damaged tissues, enhance angiogenesis, and modulate the immune response, thereby facilitating the healing process[17].

Potential Challenges and Limitations of Exosomal Cargo

While exosomes derived from mesenchymal stem cells (MSCs) hold great promise for various clinical applications, it is important to acknowledge and address the potential challenges and limitations associated with their cargo. Understanding these limitations is crucial for optimizing the therapeutic potential of exosomes and ensuring their safe and effective use in clinical settings[19].

Cargo Heterogeneity

One of the major challenges in harnessing the therapeutic potential of exosomes lies in the heterogeneity of their cargo. Exosomes contain a diverse range of molecules, including proteins, lipids, nucleic acids, and various bioactive factors. This cargo can vary depending on the cell source, culture conditions, and physiological state of the MSCs[20]. While this heterogeneity provides versatility and adaptability, it also poses challenges in terms of standardization and reproducibility[21].

The variability in cargo composition can affect the therapeutic efficacy and consistency of exosomal preparations. It is essential to identify and characterize the key functional components within exosomes to ensure their desired therapeutic effects. This requires comprehensive profiling techniques and standardized protocols for cargo analysis, which can be time-consuming and technically demanding[22].

Cargo Loading and Modification

Another challenge in utilizing exosomal cargo for therapeutic purposes is the efficient loading and modification of specific molecules. While exosomes naturally encapsulate a wide range of cargo, loading them with specific therapeutic molecules or drugs can be challenging. Strategies such as genetic engineering or chemical modification of MSCs can be employed to enhance the loading of desired cargo into exosomes. However, these approaches may introduce additional complexities and potential safety concerns[19].

Furthermore, the stability and integrity of cargo molecules within exosomes need to be carefully considered. Some cargo molecules may be susceptible to degradation or loss of activity during the isolation, purification, and storage processes. Ensuring the stability and functionality of cargo molecules within exosomes is crucial for their therapeutic efficacy[20].

Scalability and Manufacturing Challenges

The scalability and manufacturing of exosomes for clinical use present significant challenges. The current methods for exosome isolation and purification are often labor-intensive, time-consuming, and yield low quantities of exosomes. Scaling up the production of exosomes while maintaining their quality and consistency is a complex task[21].

Additionally, the manufacturing process needs to comply with good manufacturing practices (GMP) to ensure the safety, purity, and potency of exosomal products. GMP guidelines require stringent quality control measures, which can further complicate the manufacturing process and increase production costs[22].

Storage and Stability

The storage and stability of exosomes are critical considerations for their clinical applications. Exosomes are sensitive to various environmental factors, including temperature, pH, and freeze-thaw cycles. Improper storage conditions can lead to the degradation or loss of cargo molecules, compromising the therapeutic potential of exosomes[19].

Developing standardized protocols for the storage and transportation of exosomes is essential to maintain their stability and functionality. This includes optimizing storage conditions, such as temperature and cryopreservation methods, to ensure the long-term viability and efficacy of exosomal cargo[20].

Regulatory and Safety Considerations

The regulatory landscape surrounding exosome-based therapies is still evolving. As with any novel therapeutic approach, there are regulatory challenges and safety considerations that need to be addressed. The development and commercialization of exosome-based products require compliance with regulatory guidelines and approval from regulatory authorities[21].

Ensuring the safety of exosomal cargo is of utmost importance. While exosomes derived from MSCs are generally considered safe, there is a need for comprehensive safety assessments to evaluate potential risks, such as immunogenicity, off-target effects, and long-term safety profiles. Robust preclinical and clinical studies are necessary to establish the safety and efficacy of exosome-based therapies[22].

Clinical Translation and Standardization

Translating exosome-based therapies from the laboratory to clinical practice poses significant challenges. The clinical translation of exosomes requires rigorous validation, standardization, and optimization of manufacturing processes. Establishing standardized protocols for exosome isolation, characterization, and quality control is crucial for ensuring consistent and reproducible therapeutic outcomes[19].

Collaboration between researchers, clinicians, and regulatory authorities is essential to overcome these challenges and facilitate the clinical translation of exosome-based therapies. By addressing the potential limitations and developing strategies to overcome them, the therapeutic potential of exosomal cargo can be fully harnessed for the benefit of patients[20].

In conclusion, while exosomes derived from MSCs offer exciting possibilities for various clinical applications, there are several challenges and limitations that need to be addressed. These include cargo heterogeneity, efficient loading and modification of cargo, scalability and manufacturing challenges, storage and stability considerations, regulatory and safety concerns, and the need for clinical translation and standardization. By addressing these challenges, researchers and clinicians can unlock the full potential of exosomal cargo and pave the way for safe and effective exosome-based therapies[21,22].

CLINICAL APPLICATIONS OF MESENCHYMAL STEM CELL-DERIVED EXOSOMES

Exosomes in Regenerative Medicine

E xosomes derived from mesenchymal stem cells (MSCs) have emerged as a promising tool in regenerative medicine. These tiny vesicles, secreted by MSCs, play a crucial role in cell-to-cell communication and have shown immense potential in various clinical applications. In this section, we will explore the therapeutic applications of MSC-derived exosomes and provide an overview of the harvesting and manufacturing processes. Understanding Exosomes Exosomes are small extracellular vesicles that are released by cells as a means of intercellular communication. They are composed of a lipid bilayer membrane and contain a diverse cargo of proteins, lipids, and nucleic acids. These cargo molecules are selectively packaged into exosomes and can be transferred to recipient cells, influencing their behavior and function[23].

Mesenchymal Stem Cells

Mesenchymal stem cells (MSCs) are a type of adult stem cell that can be found in various tissues, including bone marrow, adipose tissue, and umbilical cord. MSCs have the remarkable ability to differentiate into multiple cell types, such as bone, cartilage, and fat cells. In addition to their differentiation potential, MSCs also possess immunomodulatory and regenerative properties, making them an attractive source for exosome production[24].

Clinical Applications of MSC-Derived Exosomes

MSC-derived exosomes hold immense potential for a wide range of clinical applications in regenerative medicine. Here are some of the key areas where these exosomes are being investigated:

1. **Tissue Repair and Regeneration:** MSC-derived exosomes have shown promising results in promoting tissue repair and regeneration in various organs, including the heart, liver, and kidneys. These exosomes can stimulate the proliferation and differentiation of resident cells, enhance angiogenesis, and modulate the immune response, ultimately leading to tissue regeneration[25].

2. **Wound Healing:** Chronic wounds, such as diabetic ulcers, pose a significant challenge in healthcare. MSC-derived exosomes have demonstrated the ability to accelerate wound healing by promoting cell migration, reducing inflamma-

tion, and enhancing tissue regeneration[23].

3. **Bone and Cartilage Regeneration:** MSC-derived exosomes have been shown to enhance bone and cartilage regeneration in preclinical studies. These exosomes can stimulate the differentiation of osteoblasts and chondrocytes, the cells responsible for bone and cartilage formation, respectively[24].

4. **Neurological Disorders:** MSC-derived exosomes have shown promise in the treatment of neurological disorders, such as stroke, traumatic brain injury, and neurodegenerative diseases. These exosomes can promote neuronal survival, enhance neuroplasticity, and modulate the inflammatory response in the central nervous system[26].

5. **Immune Modulation:** MSC-derived exosomes possess immunomodulatory properties and can regulate the immune response in various diseases, including autoimmune disorders and graft-versus-host disease. These exosomes can modulate the activity of immune cells, such as T cells and macrophages, leading to immune system balance and suppression of excessive inflammation[25].

The clinical applications of MSC-derived exosomes are still in the early stages of development, and further research is needed to fully understand their therapeutic potential. However, the preliminary results are promising, and ongoing clinical trials are providing valuable insights into their safety and efficacy[26].

In the next section, we will delve into the specific role of exosomes in neurological disorders and explore their potential applications in this field.

Exosomes in Neurological Disorders

Neurological disorders encompass a wide range of conditions that affect the brain, spinal cord, and nerves. These disorders can have a significant impact on a person's quality of life and often present complex challenges for treatment. However, recent advancements in the field of regenerative medicine have shown promising potential in utilizing exosomes derived from mesenchymal stem cells (MSCs) for the treatment of neurological disorders[27].

Understanding Exosomes in Neurological Disorders

Exosomes are small extracellular vesicles that are released by various cell types, including MSCs. These tiny vesicles play a crucial role in intercellular communication by transferring bioactive molecules, such as proteins, lipids, and nucleic acids, between cells. In the context of neurological disorders, exosomes derived from MSCs have shown remarkable therapeutic potential due to their ability to modulate various cellular processes involved in neuroprotection, neuroregeneration, and immune regulation[28].

Potential Clinical Applications

The unique properties of exosomes derived from MSCs make them an attractive candidate for the treatment of various neurological disorders. These disorders include but are not limited to:

1. **Stroke**

   Stroke is a leading cause of death and disability worldwide. The use of MSC-derived exosomes has shown promising

results in preclinical studies for stroke treatment. These exosomes have been found to promote neurovascular remodeling, reduce inflammation, and enhance neuroplasticity,
leading to improved functional recovery after stroke[29].

2. Alzheimer's Disease

Alzheimer's disease is a progressive neurodegenerative disorder characterized by the accumulation of amyloid-beta
plaques and neurofibrillary tangles in the brain. MSC-derived exosomes have demonstrated the ability to reduce
amyloid-beta-induced neurotoxicity, promote neuronal survival, and enhance cognitive function in animal models of
Alzheimer's disease[30].

3. Parkinson's Disease

Parkinson's disease is a chronic neurodegenerative disorder
characterized by the loss of dopaminergic neurons in the
brain. MSC-derived exosomes have shown potential in promoting neuronal survival, reducing inflammation, and enhancing motor function in preclinical models of Parkinson's
disease[27].

4. Spinal Cord Injury

Spinal cord injury often leads to permanent neurological
deficits due to the limited regenerative capacity of the spinal
cord. MSC-derived exosomes have been shown to promote
axonal regeneration, reduce scar formation, and enhance
functional recovery in animal models of spinal cord injury[28].

5. Multiple Sclerosis

Multiple sclerosis is an autoimmune disease characterized by the destruction of myelin, the protective covering of nerve fibers in the central nervous system. MSC-derived exosomes have demonstrated immunomodulatory effects, promoting the suppression of autoimmune responses and reducing inflammation in preclinical models of multiple sclerosis[29].

Exosomes in Cardiovascular Diseases

Cardiovascular diseases (CVDs) are a leading cause of morbidity and mortality worldwide. Despite significant advancements in medical treatments, there is still a need for innovative therapeutic approaches to effectively manage and treat these conditions. In recent years, exosomes derived from mesenchymal stem cells (MSCs) have emerged as a promising tool in the field of regenerative medicine for the treatment of cardiovascular diseases[30].

Understanding Exosomes in Cardiovascular Diseases

Exosomes are small extracellular vesicles that are released by various cell types, including MSCs. These tiny vesicles play a crucial role in cell-to-cell communication and are involved in the transfer of bioactive molecules, such as proteins, lipids, and nucleic acids, between cells. In the context of cardiovascular diseases, exosomes derived from MSCs have shown great potential in promoting tissue repair, reducing inflammation, and improving overall cardiac function[27].

The Therapeutic Potential of MSC-Derived Exosomes in Cardiovascular Diseases

MSC-derived exosomes have been extensively studied for their therapeutic effects in various cardiovascular conditions, including myocardial infarction, ischemic heart disease, and heart failure. These exosomes have been shown to enhance angiogenesis, reduce oxidative stress, promote cell survival, and modulate immune responses, all of which are crucial for the recovery and regeneration of damaged cardiac tissue[28].

One of the key mechanisms through which MSC-derived exosomes exert their therapeutic effects is by transferring their cargo of bioactive molecules to recipient cells. These molecules can include growth factors, microRNAs, and other signaling molecules that can regulate cellular processes involved in tissue repair and regeneration. For example, exosomal transfer of microRNAs has been shown to regulate gene expression in recipient cells, leading to enhanced angiogenesis and reduced inflammation in the cardiovascular system[29].

Clinical Applications of MSC-Derived Exosomes in Cardiovascular Diseases

The therapeutic potential of MSC-derived exosomes in cardiovascular diseases has been demonstrated in preclinical studies and early-phase clinical trials. These studies have shown promising results, indicating that exosomes can improve cardiac function, reduce scar formation, and enhance tissue repair in patients with cardiovascular conditions[30].

In the context of myocardial infarction, MSC-derived exosomes have been shown to promote angiogenesis and reduce inflammation, leading to improved cardiac function and reduced infarct size. These effects are attributed to the transfer of pro-angiogenic factors and anti-inflammatory molecules from the exosomes to the damaged cardiac tissue[27].

Similarly, in heart failure, MSC-derived exosomes have shown the ability to enhance cardiac repair and regeneration by promoting the survival of cardiomyocytes, reducing fibrosis, and improving overall cardiac function. The transfer of specific microRNAs and growth factors from the exosomes to the recipient cells plays a crucial role in these regenerative processes[28].

Exosomes derived from mesenchymal stem cells hold great promise as a therapeutic tool for the treatment of cardiovascular diseases. Their ability to promote tissue repair, reduce inflammation, and modulate immune responses makes them an attractive option for improving cardiac function and patient outcomes. However, further research and clinical trials are needed to fully understand the mechanisms of action and optimize the therapeutic potential of MSC-derived exosomes in cardiovascular diseases. With continued advancements in exosome research, these tiny vesicles may revolutionize the field of regenerative medicine and provide new hope for patients with cardiovascular conditions[29].

Exosomes in Immune Modulation

Exosomes derived from mesenchymal stem cells (MSCs) have gained significant attention in the field of regenerative medicine due to their unique properties and potential therapeutic applications. In this section, we will explore the role of MSC-derived exosomes in immune modulation and their potential clinical applications[31].

Understanding Immune Modulation

The immune system plays a crucial role in maintaining the body's overall health and protecting it from foreign invaders such as bacteria, viruses, and other pathogens. However, in certain conditions, the immune system can become dysregulated, leading to chronic inflammation, autoimmune diseases, and tissue damage. Immune modulation refers to the process of regulating the immune response to restore balance and promote healing[32].

The Immunomodulatory Potential of MSC-Derived Exosomes

MSC-derived exosomes have emerged as promising candidates for immune modulation due to their ability to interact with various immune cells and regulate their function. These tiny vesicles, released by MSCs, carry a cargo of proteins, lipids, and nucleic acids that can influence the behavior of immune cells[33].

One of the key mechanisms by which MSC-derived exosomes modulate the immune response is through the transfer of bioactive molecules. These exosomes can transfer microRNAs, which are small RNA molecules that can regulate gene expression in recipient cells. By delivering specific microRNAs, MSC-derived exosomes can influence the behavior of immune cells, such as T cells, B cells, and dendritic cells, and promote an anti-inflammatory environment[34].

Additionally, MSC-derived exosomes can also directly interact with immune cells through surface molecules, such as adhesion molecules and ligands. These interactions can trigger signaling pathways in immune cells, leading to the suppression of pro-inflammatory responses and the promotion of immune tolerance[31].

Clinical Applications of MSC-Derived Exosomes in Immune Modulation

The immunomodulatory properties of MSC-derived exosomes have opened up exciting possibilities for their clinical applications in various immune-related disorders. Here are some of the areas where MSC-derived exosomes show promise:

1. **Autoimmune Diseases**

   Autoimmune diseases, such as rheumatoid arthritis, multiple sclerosis, and systemic lupus erythematosus, are characterized by an overactive immune response against the body's own tissues. MSC-derived exosomes have shown potential in suppressing the immune response and reducing inflammation in these conditions[32]. By delivering immunomodulatory molecules, these exosomes can help restore immune balance and alleviate disease symptoms.

2. **Organ Transplantation**

   Organ transplantation often requires the use of immunosuppressive drugs to prevent rejection. However, these drugs can have significant side effects. MSC-derived exosomes offer a potential alternative by promoting immune tolerance and reducing the need for immunosuppressive medications[33]. These exosomes can help create an environment that is more conducive to successful organ transplantation.

3. **Inflammatory Bowel Disease**

   Inflammatory bowel disease (IBD), including conditions like Crohn's disease and ulcerative colitis, is characterized by chronic inflammation in the gastrointestinal tract. MSC-derived exosomes have shown promise in reducing inflammation and promoting tissue repair in preclinical models of IBD[34]. These exosomes can modulate the immune response in the gut, leading to a reduction in inflammation and improved disease outcomes.

In conclusion, MSC-derived exosomes hold great promise in immune modulation and have the potential to revolutionize the treatment of various immune-related disorders. Their ability to interact with immune cells and regulate the immune response makes them valuable therapeutic agents. However, further research and clinical trials are needed to fully understand their mechanisms of action and establish their safety and efficacy in different clinical settings.

EXOSOME ISOLATION AND CHARACTERIZATION TECHNIQUES

Methods for Exosome Isolation

E xosomes, small extracellular vesicles secreted by various cell types including mesenchymal stem cells (MSCs), have gained significant attention in the field of regenerative medicine and therapeutic applications. These tiny vesicles play a crucial role in cell-to-cell communication and have shown immense potential in various clinical applications. However, before exosomes can be utilized for therapeutic purposes, they need to be isolated and characterized effectively. In this section, we will explore the different methods used for exosome isolation, highlighting their advantages and limitations.

1. Ultracentrifugation

Ultracentrifugation is one of the most commonly employed

methods for exosome isolation. This technique relies on the differential centrifugation of biological samples to separate exosomes from other cellular debris and contaminants. The process involves a series of centrifugation steps at increasing speeds to pellet the exosomes. The isolated exosomes can then be further purified using sucrose density gradient centrifugation. Ultracentrifugation is a widely used method due to its simplicity and effectiveness in isolating exosomes. However, it has some limitations. The process is time-consuming and requires specialized equipment, such as an ultracentrifuge, which may not be readily available in all research or clinical settings. Additionally, ultracentrifugation may result in co-isolation of non-exosomal particles, such as lipoproteins, which can affect the purity of the exosome preparation.[35,36]

2. Size Exclusion Chromatography (SEC)

Size Exclusion Chromatography (SEC) is another commonly employed method for exosome isolation. This technique separates exosomes based on their size using a porous stationary phase. The smaller exosomes elute later, while larger particles elute earlier. SEC is advantageous as it allows for the isolation of intact exosomes without the need for ultracentrifugation. It also provides a high degree of purity and can be easily scaled up for large-scale exosome isolation. However, SEC also has its limitations. The technique may not effectively separate exosomes from other similarly sized particles, such as lipoproteins or protein aggregates. Additionally, the process can be time-consuming, and the yield of exosomes may be lower compared to other isolation methods.[37,38]

3. Precipitation methods

Precipitation methods such as polyethylene glycol (PEG) or other polymer-based precipitation, offer a simple and rapid approach for exosome isolation. These methods involve the addition of a polymer to the biological sample, which leads to the precipitation of exosomes. The precipitated exosomes can then be collected by centrifugation. Precipitation methods are advantageous as they are relatively quick and do not require specialized equipment. They also yield a high concentration of exosomes. However, these methods may result in co-precipitation of contaminants, affecting the purity of the exosome preparation. Additionally, the precipitation process may alter the exosome structure and composition, potentially impacting their biological activity.[39,40]

4. Immunoaffinity capture

Immunoaffinity capture utilizes specific antibodies or antibody-coated beads to selectively capture exosomes expressing specific surface markers. This method allows for the isolation of exosomes based on their surface protein composition. By targeting specific markers, researchers can isolate exosomes derived from a particular cell type or with specific functional properties. Immunoaffinity capture provides a highly specific and targeted approach for exosome isolation. It allows for the enrichment of exosomes expressing specific surface markers of interest. However, this method may result in the loss of exosomes that do not express the targeted markers, potentially leading to a biased exosome population.[41,42]

5. Microfluidics

Microfluidics-based techniques have emerged as promising methods for exosome isolation. These techniques utilize microfluidic devices that can efficiently separate exosomes based on their size, surface markers, or other physical properties. Microfluidic devices offer advantages such as high throughput, rapid isolation, and the ability to integrate multiple isolation steps into a single device. Microfluidics-based methods provide precise control over the isolation process and allow for the isolation of pure exosome populations. However, these techniques may require specialized equipment and expertise in microfluidic device fabrication and operation.[43,44]

Conclusion

The isolation of exosomes from mesenchymal stem cells (MSCs) is a critical step in harnessing their therapeutic potential. Various methods, such as ultracentrifugation, size exclusion chromatography, precipitation, immunoaffinity capture, and microfluidics, have been employed for exosome isolation. Each method has its advantages and limitations, and the choice of isolation method depends on the specific research or clinical requirements. It is important to note that the isolation method should be carefully selected to ensure the purity, integrity, and functionality of the isolated exosomes. Additionally, standardized protocols and quality control measures should be implemented to ensure reproducibility and comparability of exosome preparations. In the next section, we will delve into the characterization techniques used to assess the quality and content of isolated exosomes.

Characterization of Exosomes

Characterization of exosomes involves a comprehensive analysis of their physical, biochemical, and functional properties. This process allows researchers to understand the composition and behavior of exosomes, which is crucial for determining their suitability for specific clinical applications. Here, we will explore the various techniques used to characterize exosomes derived from MSCs.

1. **Physical Characterization**

   Physical characterization techniques provide valuable information about the size, shape, and concentration of exosomes. One commonly used method is nanoparticle tracking analysis (NTA), which utilizes laser light scattering to measure the size distribution and concentration of exosomes in a sample. NTA provides real-time data and can detect particles as small as 30 nanometers, making it ideal for exosome analysis[45]. Another technique used for physical characterization is electron microscopy (EM). Transmission electron microscopy (TEM) and scanning electron microscopy (SEM) allow researchers to visualize the morphology and structure of exosomes at high resolution. These techniques provide valuable insights into the size, shape, and surface characteristics of exosomes[46].

2. Biochemical Characterization

Biochemical characterization techniques focus on identifying the specific molecules present in exosomes. One widely used method is Western blotting, which involves separating exosomal proteins using gel electrophoresis and then detecting specific proteins using antibodies. This technique allows researchers to identify protein markers that are commonly found in exosomes, such as CD9, CD63, and CD81[47]. Flow cytometry is another valuable tool for exosome characterization. By labeling exosomes with fluorescent antibodies, researchers can analyze the expression of specific surface markers. This technique provides quantitative data on the presence and abundance of specific proteins on the exosomal surface[48].

3. Functional Characterization

Functional characterization of exosomes involves assessing their biological activity and potential therapeutic effects. One common approach is to evaluate the ability of exosomes to transfer their cargo to recipient cells. This can be done by labeling exosomes with fluorescent dyes or genetic markers and tracking their uptake by recipient cells using fluorescence microscopy or flow cytometry[45]. Additionally, functional assays can be performed to assess the impact of exosomes on various cellular processes. For example, researchers can investigate the ability of exosomes to promote cell proliferation, migration, or differentiation. These functional assays provide insights into the potential therapeutic effects of exosomes derived from MSCs[48].

Furthermore, it is essential to establish guidelines for the characterization of exosomes, including the minimum requirements for physical, biochemical, and functional analysis. These guidelines will help researchers and clinicians compare and interpret exosome data from different studies, facilitating the advancement of exosome-based therapies[46].

Quality Control and Standardization of Exosome Products

Exosomes derived from mesenchymal stem cells (MSCs) hold immense potential for clinical applications due to their unique properties and therapeutic capabilities. However, to ensure their safety and efficacy, it is crucial to establish robust quality control measures and standardization protocols for exosome products. This section will delve into the importance of quality control and standardization in exosome manufacturing, highlighting the challenges and strategies involved in maintaining consistent and reliable exosome products[49].

Importance of Quality Control

Quality control is a critical aspect of exosome production as it ensures that the final product meets the required standards of safety, purity, and potency. By implementing stringent quality control measures, researchers and manufacturers can minimize batch-to-batch variability and ensure reproducibility, thereby enhancing the reliability and effectiveness of exosome-based therapies[50].

One of the primary goals of quality control is to assess the physical and chemical characteristics of exosomes. This includes evaluating their size, morphology, concentration, surface markers, and cargo content. By thoroughly characterizing exosomes, researchers can gain insights into their biological properties and establish a foundation for standardization[51].

Standardization of Exosome Products

Standardization plays a crucial role in the development and commercialization of exosome-based therapies. It involves defining and implementing consistent manufacturing processes, quality control protocols, and product specifications to ensure uniformity and reproducibility across different batches and manufacturers[52].

One of the key challenges in standardizing exosome products is the lack of consensus regarding optimal isolation and characterization techniques. Different isolation methods, such as ultracentrifugation, size exclusion chromatography, and precipitation-based methods, can yield exosomes with varying purity and yield. Therefore, it is essential to establish standardized protocols for isolation to ensure consistent product quality[49].

Similarly, the characterization of exosomes requires standardized techniques to assess their size, concentration, surface markers, and cargo content. Techniques such as nanoparticle tracking analysis (NTA), transmission electron microscopy (TEM), flow cytometry, and proteomic analysis are commonly employed for exosome characterization. Standardizing these techniques and establishing reference materials can facilitate accurate and reliable characterization of exosome products[50].

Challenges in Quality Control and Standardization

Several challenges need to be addressed to achieve effective quality control and standardization of exosome products. One of the primary challenges is the heterogeneity of exosomes. Exosomes derived from different cell sources or under varying conditions can exhibit variations in size, cargo content, and functional properties. Therefore, it is crucial to establish standardized protocols that account for these variations and ensure consistent product quality[51].

Another challenge is the lack of standardized reference materials and assays for exosome characterization. Without well-defined reference materials, it becomes challenging to compare and validate the results obtained from different laboratories or manufacturers. Efforts are underway to develop reference materials and establish standardized assays for exosome characterization, which will greatly facilitate quality control and standardization[52].

Strategies for Quality Control and Standardization

To overcome the challenges associated with quality control and standardization, several strategies can be employed. Collaboration among researchers, manufacturers, and regulatory authorities is essential to establish consensus on standardized protocols and guidelines for exosome production and characterization. This collaboration can help in sharing knowledge, best practices, and reference materials, thereby promoting uniformity and reproducibility in exosome manufacturing[49].

The use of advanced analytical techniques and technologies can also aid in quality control and standardization. For instance, the development of high-resolution imaging techniques, such as cryo-electron microscopy, can provide detailed insights into the morphology and structure of exosomes. Similarly, the use of mass spectrometry-based proteomic analysis can enable comprehensive profiling of exosomal cargo, facilitating standardized characterization[50].

Furthermore, the implementation of good manufacturing practices (GMP) is crucial for ensuring consistent product quality and safety. GMP guidelines provide a framework for maintaining quality control throughout the manufacturing process, including facility design, personnel training, documentation, and quality assurance. Adhering to GMP guidelines can help minimize the risk of contamination, ensure batch-to-batch consistency, and meet regulatory requirements[51].

Quality control and standardization are vital for the successful translation of exosome-based therapies into clinical applications. By establishing robust quality control measures and standardized protocols, researchers and manufacturers can ensure the safety, efficacy, and reproducibility of exosome products. Collaboration among stakeholders, the development of reference materials, and the use of advanced analytical techniques are key strategies in achieving effective quality control and standardization. As the field of exosome research continues to evolve, it is imperative to prioritize these efforts to unlock the full potential of exosomal elixir in revolutionizing regenerative medicine and therapeutic interventions[52].

Scaling Up Exosome Production

Exosomes derived from mesenchymal stem cells (MSCs) hold immense potential for various clinical applications. As the demand for exosome-based therapies continues to grow, it becomes crucial to develop efficient methods for scaling up exosome production. This section will explore the challenges and strategies involved in scaling up the manufacturing process of MSC-derived exosomes[53].

Challenges in Scaling Up Exosome Production

Scaling up the production of MSC-derived exosomes presents several challenges that need to be addressed to meet the increasing demand for these therapeutic agents. Some of the key challenges include:

1. **Yield and Efficiency**

   One of the primary challenges in scaling up exosome production is achieving a high yield and maintaining the efficiency of the process. MSCs naturally release exosomes, but the quantity obtained from a single cell source is limited. To overcome this limitation, researchers are exploring different strategies to enhance exosome production, such as optimizing culture conditions, using bioreactors, and genetic engineering of MSCs[53].

2. **Standardization and Quality Control**

   Maintaining consistent quality and ensuring the reproducibility of exosome production is essential for clinical applications. However, scaling up the production process can introduce variability, making it challenging to achieve standardized exosome products. Implementing robust quality control measures and standardizing manufacturing protocols are crucial to ensure the safety and efficacy of exosome-based therapies[53].

3. **Purification and Isolation**

   Obtaining pure and highly concentrated exosome preparations is critical for their therapeutic efficacy. Scaling up the production process can introduce additional impurities and contaminants, making the purification and isolation of exosomes more challenging. Developing efficient and scalable methods for exosome purification is essential to obtain high-quality exosome products[53].

4. Cost and Scalability

Scaling up the production of exosomes can significantly impact the overall cost of manufacturing. The development of cost-effective and scalable production methods is necessary to make exosome-based therapies more accessible and affordable for patients. Exploring alternative sources of MSCs and optimizing culture conditions can help reduce production costs and increase scalability[53].

Strategies for Scaling Up Exosome Production

To overcome the challenges associated with scaling up exosome production, researchers are exploring various strategies and technologies. Some of the key strategies include:

1. Bioreactor Systems

Bioreactor systems provide a controlled environment for the expansion and differentiation of MSCs, allowing for large-scale production of exosomes. These systems offer advantages such as increased cell yield, improved reproducibility, and scalability. Bioreactors can be designed to mimic the physiological conditions required for optimal exosome production, enhancing the efficiency of the manufacturing process[53].

2. Genetic Engineering of MSCs

Genetic engineering approaches can be employed to enhance the production and secretion of exosomes from MSCs. By modifying the genetic makeup of MSCs, researchers can increase the yield and potency of exosomes. Genetic engineering techniques can also be used to introduce specific therapeutic cargo into exosomes, further enhancing their therapeutic potential[53].

3. **Microcarrier-Based Culture Systems**

Microcarrier-based culture systems provide a three-dimensional environment for MSC growth, allowing for higher cell densities and increased exosome production. These systems offer advantages such as improved scalability, enhanced cell-to-cell interactions, and increased surface area for cell attachment. Microcarrier-based culture systems can be integrated with bioreactors to facilitate large-scale exosome production[54].

4. **Process Optimization and Automation**

Optimizing the manufacturing process and automating critical steps can significantly improve the scalability and efficiency of exosome production. By carefully analyzing and optimizing each step, researchers can identify bottlenecks and implement strategies to streamline the process. Automation of key steps, such as cell expansion, exosome isolation, and purification, can reduce human error and increase production throughput[53].

Conclusion

Scaling up the production of MSC-derived exosomes is a critical step towards realizing their full potential in clinical applications. Overcoming the challenges associated with scaling up exosome production requires a multidisciplinary approach, involving advancements in bioreactor systems, genetic engineering, and process optimization. By addressing these challenges and implementing scalable manufacturing strategies, we can meet the increasing demand for exosome-based therapies and unlock their healing potential for a wide range of diseases and conditions[53].

Manufacturing and Quality Control of Exosome-Based Therapies

Good Manufacturing Practices (GMP) for Exosome Production

Exosomes derived from mesenchymal stem cells (MSCs) hold immense potential for clinical applications in regenerative medicine, neurological disorders, cardiovascular diseases, and immune modulation. However, to ensure the safety, efficacy, and quality of exosome-based therapies, it is crucial to adhere to Good Manufacturing Practices (GMP) during the production process. GMP guidelines provide a framework for the consistent and controlled manufacturing of exosomes, ensuring that they meet the required standards for clinical use.[55]

Understanding Good Manufacturing Practices (GMP)

Good Manufacturing Practices (GMP) are a set of guidelines and regulations established by regulatory authorities, such as the Food and Drug Administration (FDA) in the United States and the European Medicines Agency (EMA) in Europe. These guidelines outline the minimum requirements for the production, quality control, and documentation of pharmaceutical products, including exosome-based t herapies.[56]

GMP guidelines are designed to ensure that the manufacturing process is well-defined, controlled, and reproducible, minimizing the risk of contamination, errors, and variations in product quality. By adhering to GMP, manufacturers can consistently produce exosomes that are safe, pure, potent, and of high quality.

GMP Requirements for Exosome Production

To comply with GMP guidelines, manufacturers of exosome-based therapies must implement several key requirements throughout the production process. These requirements include:

1. **Facility Design and Control**

   GMP-compliant facilities must be designed and maintained to minimize the risk of contamination and ensure a controlled environment for exosome production. This includes appropriate air filtration systems, temperature and humidity control, and segregation of different production areas to prevent cross-contamination.

2. **Equipment and Instrumentation**

   Manufacturers must use validated equipment and instrumentation that are regularly calibrated and maintained to ensure accuracy and reliability. This includes specialized equipment for exosome isolation, purification, and characterization, such as ultracentrifuges, filtration systems, and particle analyzers.

3. **Raw Material Control**

GMP requires strict control and documentation of all raw materials used in the production of exosomes. This includes the use of qualified suppliers, proper storage conditions, and thorough testing to ensure the identity, purity, and quality of raw materials. Any deviations or changes in raw materials must be carefully evaluated and documented.

4. **Standard Operating Procedures (SOPs)**

Manufacturers must develop and implement detailed Standard Operating Procedures (SOPs) for all critical steps in the production process. SOPs provide step-by-step instructions for personnel, ensuring consistency and reproducibility. These procedures cover aspects such as cell culture, exosome isolation, purification, characterization, and quality control testing.[57]

5. **Personnel Training and Qualification**

GMP emphasizes the importance of well-trained and qualified personnel involved in the production of exosomes. Manufacturers must provide comprehensive training programs to ensure that personnel understand and follow the established procedures. Training should cover topics such as aseptic techniques, equipment operation, documentation practices, and safety protocols.

6. **Quality Control and Testing**

   GMP requires rigorous quality control and testing throughout the production process to ensure the safety, purity, and
   potency of exosomes. This includes testing of raw materials, in-process samples, and final exosome products. Quality
   control tests may include assessments of particle size, concentration, surface markers, genetic material, and functional
   activity of exosomes.

7. **Documentation and Record-Keeping**

   Accurate and comprehensive documentation is a funda-
   mental requirement of GMP. Manufacturers must maintain
   detailed records of all aspects of the production process, including batch records, equipment logs, testing results, and
   deviations from established procedures. These records serve
   as evidence of compliance and provide traceability for each
   batch of exosomes.

Benefits of GMP Compliance in Exosome Production

Adhering to GMP guidelines for exosome production offers several
benefits, including:

1. Ensuring the safety and efficacy of exosome-based therapies
   for patients.

2. Minimizing the risk of contamination, errors, and variations
   in product quality.

3. Facilitating regulatory approval and compliance with regulatory authorities.

4. Enhancing the reputation and credibility of manufacturers in the field of exosome-based therapies.

5. Promoting consistency, reproducibility, and scalability of exosome production processes.

Good Manufacturing Practices (GMP) play a vital role in the production of exosome-based therapies derived from mesenchymal stem cells. By following GMP guidelines, manufacturers can ensure the safety, efficacy, and quality of exosomes, thereby maximizing their therapeutic potential. Adhering to GMP requirements, including facility design and control, equipment validation, raw material control, SOPs, personnel training, quality control, and documentation, is essential for the successful translation of exosome research into clinical applications. GMP compliance not only benefits patients but also contributes to the advancement and commercialization of exosome therapies.

Safety Considerations and Risk Assessment in Exosome Manufacturing

Ensuring the safety and minimizing the risks associated with exosome manufacturing is of paramount importance in the development of exosome-based therapies. As these therapies progress towards clinical applications, it is crucial to establish robust safety considerations and conduct thorough risk assessments to safeguard patient well-being. In this section, we will explore the key safety considerations and risk assessment strategies involved in exosome manufacturing.

Source of Mesenchymal Stem Cells (MSCs): The selection of an appropriate source for MSCs is critical to ensure the safety and efficacy of exosome-based therapies. MSCs can be derived from various tissues, such as bone marrow, adipose tissue, and umbilical cord. It is essential to thoroughly characterize the MSCs and ensure their quality, purity, and absence of any potential contaminants or pathogens[58].

Donor Screening and Testing: Donor screening and testing are essential steps to minimize the risk of transmitting infectious diseases through exosome-based therapies. Donors should undergo rigorous screening processes, including medical history evaluation, physical examination, and serological testing for infectious diseases. These measures help identify potential risks and ensure the safety of the derived exosomes[59].

Exosome Isolation and Purification: The isolation and purification processes play a crucial role in maintaining the safety and integrity of exosomes. It is important to employ validated and standardized methods for exosome isolation to minimize the risk of contamination and maintain the purity of the final product. Quality control measures, such as sterility testing, endotoxin testing, and mycoplasma testing, should be implemented to ensure the absence of harmful contaminants[60].

Characterization of Exosomes: Thorough characterization of exosomes is essential to assess their safety and efficacy. This includes evaluating their size, morphology, surface markers, and cargo content. Techniques such as transmission electron microscopy (TEM), nanoparticle tracking analysis (NTA), and flow cytometry can be employed to characterize exosomes and ensure their quality and consistency[61].

Risk Assessment in Exosome Manufacturing

Risk Identification: The first step in risk assessment is identifying potential risks associated with exosome manufacturing. This involves evaluating the entire manufacturing process, from the isolation of MSCs to the final formulation of exosome-based therapies. Potential risks may include contamination, variability in exosome composition, inadequate characterization, and potential immunogenicity.

Risk Analysis: Once the risks are identified, a comprehensive risk analysis should be conducted to assess the severity and likelihood of each risk. This involves evaluating the potential impact on patient safety and the probability of occurrence. Risk analysis helps prioritize risks and determine the necessary mitigation strategies.

Risk Mitigation: Risk mitigation strategies aim to minimize or eliminate identified risks. This may involve implementing process controls, such as stringent aseptic techniques, to prevent contamination. Standardizing manufacturing processes and employing validated analytical methods can help reduce variability and ensure consistent product quality. Additionally, implementing appropriate storage and transportation conditions can mitigate the risk of product degradation.

Risk Monitoring and Management: Continuous monitoring and management of risks throughout the manufacturing process are essential to maintain product safety. Regular audits, inspections, and quality control assessments should be conducted to identify any deviations or potential risks. Implementing corrective and preventive actions (CAPAs) in response to identified risks helps ensure ongoing safety and quality.

Regulatory Guidelines for Exosome-Based Therapies

Regulatory authorities play a crucial role in ensuring the safety and efficacy of exosome-based therapies. As these therapies progress towards clinical translation, regulatory guidelines provide a framework for manufacturers to adhere to. Regulatory agencies, such as the U .S. Food and Drug Administration (FDA) and the European Medicines Agency (EMA), have specific requirements for the development, manufacturing, and clinical use of exosome-based therapies. These guidelines encompass safety considerations, risk assessment, quality control, and documentation requirements.

Manufacturers of exosome-based therapies must comply with Good Manufacturing Practices (GMP) to ensure the safety, quality, and consistency of their products. GMP guidelines outline the necessary infrastructure, equipment, personnel training, and documentation practices required for exosome manufacturing. Adhering to these guidelines helps minimize risks, maintain product integrity, and ensure patient safety.

In conclusion, safety considerations and risk assessment are crucial aspects of exosome manufacturing. Thorough donor screening, standardized isolation and purification processes, robust quality control measures, and adherence to regulatory guidelines are essential to ensure the safety and efficacy of exosome-based therapies. By implementing these measures, manufacturers can mitigate risks, maintain product quality, and pave the way for the successful clinical translation of exosome therapies.

Regulatory Agencies and Frameworks

Regulatory oversight of exosome-based therapies varies across different countries and regions. In the United States, the Food and Drug Administration (FDA) plays a crucial role in regulating the development, manufacturing, and clinical use of these therapies. The FDA classifies exosome-based products as biological products and evaluates them under the same regulatory framework as other cellular and gene therapies[58].

Similarly, the European Medicines Agency (EMA) oversees the regulation of exosome-based therapies in the European Union. The EMA provides guidelines and requirements for the development, manufacturing, and clinical evaluation of these therapies, ensuring their safety and efficacy[59].

Other countries, such as Japan, Australia, and Canada, also have their own regulatory agencies and frameworks for the evaluation and approval of exosome-based therapies. It is essential for researchers and developers to familiarize themselves with the specific regulations and guidelines of the respective countries where they intend to conduct clinical trials or seek regulatory approval[60].

Preclinical Studies and Investigational New Drug (IND) Application

Before initiating clinical trials, exosome-based therapies must undergo rigorous preclinical studies to establish their safety and efficacy. These studies involve in vitro experiments, animal models, and toxicity assessments to evaluate the therapeutic potential and potential risks associated with the therapy[61].

Once preclinical studies demonstrate promising results, researchers can submit an Investigational New Drug (IND) application to the regulatory agency, such as the FDA in the United States. The IND application provides comprehensive data on the therapy, including preclinical data, manufacturing processes, and proposed clinical trial protocols. The regulatory agency reviews the IND application to assess the safety and feasibility of proceeding with clinical trials.

Clinical Trials and Phases

Clinical trials are essential for evaluating the safety and efficacy of exosome-based therapies in humans. These trials are typically conducted in multiple phases, each with specific objectives and requirements.

Phase 1 trials involve a small number of healthy volunteers or patients and primarily focus on assessing the safety, dosage, and potential side effects of the therapy. Phase 2 trials expand the study population to a larger group of patients and aim to gather preliminary data on the therapy's effectiveness and optimal dosage. Phase 3 trials involve a larger patient population and compare the therapy's efficacy against existing standard treatments or placebos. These trials provide critical evidence of the therapy's effectiveness and safety profile, which is crucial for regulatory approval.

Regulatory Approval and Post-Marketing Surveillance

After successful completion of clinical trials, exosome-based therapies can seek regulatory approval for commercialization. The regulatory agency carefully evaluates the clinical trial data, manufacturing processes, and safety profiles to determine whether the therapy meets the necessary standards for approval. Once approved, exosome-based therapies are subject to post-marketing surveillance to monitor their long-term safety and effectiveness in real-world settings. This surveillance helps identify any rare or long-term side effects that may not have been apparent during clinical trials.

Manufacturing and Quality Control Standards

Regulatory guidelines also emphasize the importance of adhering to Good Manufacturing Practices (GMP) for the production of exosome-based therapies. GMP ensures that the manufacturing processes are consistent, controlled, and meet the required quality standards.

These guidelines cover various aspects, including facility design, personnel training, documentation, quality control, and product labeling.

Quality control measures are crucial throughout the manufacturing process to ensure the safety, purity, and potency of exosome-based therapies. These measures include rigorous testing of raw materials, characterization of the final product, and monitoring of critical quality attributes. Quality control also extends to storage, transportation, and distribution of the therapy to maintain its integrity and efficacy.

International Harmonization and Standardization

Efforts are underway to harmonize and standardize the regulatory guidelines for exosome-based therapies globally. Organizations such as the International Society for Extracellular Vesicles (ISEV) and the International Organization for Standardization (ISO) are actively working towards establishing consensus on terminology, characterization methods, and quality control standards for exosome research and therapy development.International harmonization and standardization not only facilitate regulatory processes but also promote collaboration, knowledge sharing, and the advancement of exosome-based therapies on a global scale.

In conclusion, regulatory guidelines play a crucial role in ensuring the safety, efficacy, and quality control of exosome-based therapies. Adherence to these guidelines is essential for the successful development, clinical translation, and commercialization of these therapies. As the field continues to evolve, international harmonization and standardization efforts will further streamline the regulatory processes and accelerate the realization of the therapeutic potential of exosomes derived from mesenchymal stem cells.

CONCLUSION AND FUTURE PERSPECTIVES

Summary of Key Findings and Insights

In this book, we have explored the fascinating world of exosomes derived from mesenchymal stem cells (MSCs) and their immense potential in clinical applications. Throughout the chapters, we have delved into the composition of exosomal cargo, the role of exosomes in cell communication and tissue regeneration, and the various clinical applications of MSC-derived exosomes. We have also discussed the challenges and opportunities in exosome-based therapies, the techniques for exosome isolation and characterization, and the manufacturing and quality control processes involved. In this section, we will summarize the key findings and insights from our exploration and discuss the future perspectives of exosome research and clinical applications.

Exosomes, small extracellular vesicles secreted by cells, play a crucial role in intercellular communication and have emerged as promising therapeutic agents. MSC-derived exosomes, in particular, have gained significant attention due to their unique properties and potential clinical applications. These exosomes are released by MSCs, a type of adult stem cell found in various tissues, including bone marrow, adipose tissue, and umbilical cord. MSCs have the ability to differentiate into multiple cell types and possess immunomodulatory and regenerative properties.

One of the key findings of our exploration is the diverse range of clinical applications of MSC-derived exosomes. These exosomes have shown promising results in regenerative medicine, neurological disorders, cardiovascular diseases, and immune modulation. In regenerative medicine, MSC-derived exosomes have demonstrated their ability to promote tissue repair and regeneration by stimulating cell proliferation, angiogenesis, and extracellular matrix remodeling. They have shown potential in the treatment of conditions such as osteoarthritis, wound healing, and tissue damage caused by ischemia.

In the field of neurological disorders, MSC-derived exosomes have shown neuroprotective effects and the ability to enhance neuronal survival and regeneration. They have shown promise in the treatment of conditions such as stroke, traumatic brain injury, and neurodegenerative diseases like Alzheimer's and Parkinson's. The ability of MSC-derived exosomes to cross the blood-brain barrier and deliver therapeutic cargo to the central nervous system makes them an attractive option for targeted drug delivery in neurological disorders.

Cardiovascular diseases, including myocardial infarction and heart failure, pose a significant burden on global health. MSC-derived exosomes have shown potential in promoting cardiac repair and regeneration by enhancing angiogenesis, reducing inflammation, and improving cardiac function. These exosomes have been found to stimulate the proliferation and migration of endothelial cells and promote the survival of cardiomyocytes. They hold promise as a non-invasive therapeutic approach for cardiovascular diseases.

Another significant finding is the immunomodulatory properties of MSC-derived exosomes. These exosomes have the ability to modulate immune responses by suppressing inflammation and regulating the activity of immune cells. They have shown potential in the treatment of autoimmune diseases, graft-versus-host disease, and organ transplantation. MSC-derived exosomes can promote immune tolerance and reduce the risk of immune rejection, making them a valuable tool in immune modulation therapies.

The harvesting and manufacturing processes of MSC-derived exosomes are crucial for their clinical translation. Various methods for exosome isolation have been developed, including ultracentrifugation, size exclusion chromatography, and precipitation-based techniques. These methods allow for the purification of exosomes from MSC culture supernatants or other sources. Characterization techniques such as electron microscopy, nanoparticle tracking analysis, and proteomic analysis help in confirming the presence of exosomes and assessing their size, morphology, and cargo.

Manufacturing exosome-based therapies involves scaling up the production of MSCs and optimizing the exosome isolation process. Good Manufacturing Practices (GMP) ensure the quality and safety of exosome products for clinical use. Quality control measures, including testing for sterility, endotoxin levels, and identity verification, are essential to ensure the consistency and efficacy of exosome-based therapies. Regulatory guidelines play a crucial role in the development and approval of exosome-based therapies, ensuring their safety and efficacy.

In conclusion, the field of exosome research and clinical applications is rapidly evolving, and MSC-derived exosomes hold immense promise as therapeutic agents. The key findings and insights from our exploration highlight the diverse range of clinical applications of MSC-derived exosomes, including regenerative medicine, neurological disorders, cardiovascular diseases, and immune modulation. The harvesting and manufacturing processes of MSC-derived exosomes are crucial for their clinical translation, and adherence to regulatory guidelines and quality control measures is essential. The future of exosome research and clinical applications looks promising, with emerging trends and opportunities for further advancements. The potential of exosomal elixir, derived from MSCs, to unleash the healing potential of exosomes opens up new possibilities for the treatment of various diseases and injuries. As we continue to unravel the mysteries of exosomes, we are hopeful that they will revolutionize the field of medicine and bring new hope to patients worldwide.

Final Thoughts and Closing Remarks

In this book, we have explored the fascinating world of exosomes and their potential in revolutionizing the field of regenerative medicine. Exosomes, particularly those derived from mesenchymal stem cells (MSCs), have emerged as promising therapeutic agents due to their unique properties and ability to communicate with target cells. As we conclude our journey through the realm of exosomes, let us reflect on the key findings and insights we have gained, and contemplate the future directions and implications of this groundbreaking research.

Throughout this book, we have delved into the intricate nature of exosomes and their role in cell communication and tissue regeneration. Exosomes are small vesicles secreted by cells, including MSCs, that contain a diverse cargo of proteins, nucleic acids, and lipids. These tiny messengers play a crucial role in intercellular communication, facilitating the transfer of biomolecules between cells and influencing various physiological and pathological processes.

One of the most exciting aspects of exosomes is their potential in clinical applications. Exosomes derived from MSCs have shown immense promise in regenerative medicine, neurological disorders, cardiovascular diseases, and immune modulation. These tiny vesicles possess regenerative properties that can stimulate tissue repair and regeneration, making them an attractive alternative to traditional cell-based therapies. Furthermore, their ability to cross the blood-brain barrier and modulate immune responses has opened up new avenues for the treatment of neurological disorders and immune-related conditions.

To harness the therapeutic potential of exosomes, it is crucial to understand the methods of exosome harvesting and manufacturing. The isolation of exosomes from MSCs involves a series of steps, including cell culture, collection of conditioned media, and subsequent purification using various techniques such as ultracentrifugation, size exclusion chromatography, and precipitation methods. These methods allow for the separation of exosomes from other cellular debris and contaminants, ensuring the purity and integrity of the final exosome product.

Manufacturing exosomes for clinical use requires adherence to strict quality control measures and regulatory guidelines. Good Manufacturing Practices (GMP) ensure that exosome production follows standardized protocols and meets the required safety and quality standards. Quality control measures, such as assessing the size, concentration, and cargo content of exosomes, are essential to ensure consistency and reproducibility of the therapeutic product. Additionally, safety considerations and risk assessments are crucial to minimize any potential adverse effects associated with exosome-based therapies.

As we look towards the future, it is evident that exosome research holds immense promise. The field of exosome-based therapies is rapidly evolving, with ongoing clinical trials exploring the efficacy and safety of exosome treatments in various diseases. These trials will provide valuable insights into the therapeutic potential of exosomes and pave the way for their widespread clinical adoption.

However, it is important to acknowledge the challenges and limitations that accompany exosome research. The heterogeneity of exosomes, both in terms of their cargo composition and functional properties, poses challenges in standardizing and scaling up exosome production. Furthermore, the regulatory landscape surrounding exosome-based therapies is still evolving, necessitating the establishment of clear guidelines and frameworks to ensure their safe and ethical use.

In conclusion, the potential of exosomes, particularly those derived from MSCs, in regenerative medicine and therapeutic applications is immense. These tiny vesicles hold the key to unlocking the body's natural healing potential and offer a promising alternative to traditional cell-based therapies. However, further research is needed to fully understand the mechanisms of action, optimize manufacturing processes, and establish robust regulatory frameworks.

As we bid farewell to this book, we hope that it has provided you with a comprehensive understanding of exosomes and their role in unleashing the healing potential of mesenchymal stem cell-derived exosomes. The future of exosome research is bright, and we eagerly anticipate the advancements and breakthroughs that lie ahead. Let us embrace the promise of exosomal elixir and its potential to transform the landscape of regenerative medicine.

REFERENCES

1. Lu, Y., Mai, Z., Cui, L., & Zhao, X. (2023). Engineering exosomes and biomaterial-assisted exosomes as therapeutic carriers for bone regeneration. *Stem Cell Research & Therapy, 14*(1), 1-16. https://doi.org/10.1186/s13287-023-03275-x

2. Rezabakhsh, A., Sokullu, E., & Rahbarghazi, R. (2021). Applications, challenges and prospects of mesenchymal stem cell exosomes in regenerative medicine. *Stem Cell Research & Therapy, 12*(1), 1-14. https://doi.org/10.1186/s13287-021-02596-z

3. Popowski, K. D., Lutz, H., Hu, S., George, A., Dinh, P., & Cheng, K. (2020). Exosome therapeutics for lung regenerative medicine. *Journal of Extracellular Vesicles, 9*(1), 1785161. https://doi.org/10.1080/20013078.2020.1785161

4. Hade, M. D., Suire, C. N., & Suo, Z. (2021). Mesenchymal

Stem Cell-Derived Exosomes: Applications in Regenerative Medicine. *Cells, 10*(8), 1959. https://doi.org/10.3390/cells 10081959

5. Ma, Z., Yang, J., Lu, Y., Liu, Z., & Wang, X. (2020). Mesenchymal stem cell-derived exosomes: Toward cell-free therapeutic strategies in regenerative medicine. *World Journal of Stem Cells, 12*(8), 814. https://doi.org/10.4252/wjsc.v12.i 8.814

6. Zhou, A., Jou, E., Lu, V., Zhang, J., Chabra, S., Abishek, J., Wong, E., Zeng, X., & Guo, B. (2023). Using Pre-Clinical Studies to Explore the Potential Clinical Uses of Exosomes Secreted from Induced Pluripotent Stem Cell-Derived Mesenchymal Stem cells. *Tissue Engineering and Regenerative Medicine.* https://doi.org/10.1007/s13770-023-00557-6

7. Janockova, J., Slovinská, L., Harvanova, D., Spakova, T., & Rosocha, J. (2021). New therapeutic approaches of mesenchymal stem cells-derived exosomes. *Journal of Biomedical Science, 28*(1), 36. https://doi.org/10.1186/s12929-021-00 736-4

8. Hassanzadeh, A., Rahman, H., Markov, A., Endjun, J. J., Zekiy, A., Chartrand, M., Beheshtkhoo, N., Kouhbanani, M. A. J., Marofi, F., Nikoo, M., & Jarahian, M. (2021). Mesenchymal stem/stromal cell-derived exosomes in regenerative medicine and cancer; overview of development, challenges, and opportunities. *Stem Cell Research & Therapy, 12*(1), 297. https://doi.org/10.1186/s13287-021-02378-7

9. Shen, M., & Chen, T. (2021). Mesenchymal Stem Cell-De-

rived Exosomes and Their Potential Agents in Hematological Diseases. *Oxidative Medicine and Cellular Longevity*, 2021. https://doi.org/10.1155/2021/4539453

10. Su, R. (2022). Mesenchymal Stem Cell Exosomes as Nanotherapeutic Agents for Neurodegenerative Diseases. *Health Science and Technology*, 2. https://doi.org/10.54097/hset.v 2i.549

11. Guo, G., Tan, Z.-L., Liu, Y., Shi, F., & She, J. (2022). The therapeutic potential of stem cell-derived exosomes in ulcerative colitis and colorectal cancer. *Stem Cell Research & Therapy*, 13(1). https://doi.org/10.1186/s13287-022-0281 1-5

12. Yang, Z., Li, Y., & Wang, Z. (2022). Recent Advances in the Application of Mesenchymal Stem Cell-Derived Exosomes for Cardiovascular and Neurodegenerative Disease Therapies. *Pharmaceutics*, 14(3), 618. https://doi.org/10.3390/p harmaceutics14030618

13. Yao, X., Liao, B., Chen, F., Liu, L., Wu, K., Hao, Y., Li, Y., Wang, Y., Fan, R., Yin, J., Liu, L., & Guo, Y. (2023). Comparison of proteomic landscape of extracellular vesicles in pleural effusions isolated by three strategies. *Frontiers in Bioengineering and Biotechnology* https://doi.org/10.3389 /fbioe.2023.1108952

14. Chen, Y.-s., Lin, E., Chiou, T., & Harn, H. (2019). Exosomes in clinical trial and their production in compliance with good manufacturing practice. *Tzu Chi Medical Journal*, 31(2), 113–120. https://doi.org

/10.4103/tcmj.tcmj_182_19

15. Ma, Z., Yang, J., Lu, Y.-b., Liu, Z., & Wang, X. (2020). Mesenchymal stem cell-derived exosomes: Toward cell-free therapeutic strategies in regenerative medicine. *World Journal of Stem Cells*, 12(8), 814. DOI:10.4252/wjsc.v12.i8.814

16. Nikfarjam, S., Rezaie, J., Zolbanin, N. M., & Jafari, R. (2020). Mesenchymal stem cell derived-exosomes: a modern approach in translational medicine. *Journal of Translational Medicine*, 18(1), 449. DOI:10.1186/s12967-020-02622-3

17. Hade, M. D., Suire, C. N., & Suo, Z. (2021). Mesenchymal Stem Cell-Derived Exosomes: Applications in Regenerative Medicine. *Cells*, 10(8), 1959. DOI:10.3390/cells10081959

18. Rezabakhsh, A., Sokullu, E., & Rahbarghazi, R. (2021). Applications, challenges and prospects of mesenchymal stem cell exosomes in regenerative medicine. *Stem Cell Research & Therapy*, 12(1), 530. DOI:10.1186/s13287-021-02596-z

19. Janockova, J., Slovinská, L., Harvanova, D., Spakova, T., & Rosocha, J. (2021). New therapeutic approaches of mesenchymal stem cells-derived exosomes. *Journal of Biomedical Science*, 28(1), 36. https://doi.org/10.1186/s12929-021-00736-4

20. Liu, W.-z., Ma, Z., Li, J.-r., & Kang, X. (2021). Mesenchymal stem cell-derived exosomes: therapeutic opportunities and challenges for spinal cord injury. *Stem Cell Research & Therapy*, 12(1), 102. https://doi.org/10.1186/s13287-021-02153-8

21. Ahmadi,   M.,   Mahmoodi,   M.,   Shoaran,   M., Nazari-Khanamiri, F., & Rezaie, J. (2022). Harnessing Normal and Engineered Mesenchymal Stem Cells Derived Exosomes for Cancer Therapy: Opportunity and Challenges. *International Journal of Molecular Sciences, 23*(22), 13974. https://doi.org/10.3390/ijms232213974

22. Zhou, X., & Kalluri, R. (2020). The Biology and Therapeutic Potential of Mesenchymal Stem Cells derived Exosomes. *Cancer Science, 111*(9), 3100-3110. https://doi.org/10.111 1/cas.14563

23. Smolinská, V., Boháč, M., & Danišovič, Ľ. (2023). Current status of the applications of conditioned media derived from mesenchymal stem cells for regenerative medicine. *Physiological Research*. https://dx.doi.org/10.33549/physiolres.9 35186

24. Zhou, A., Jou, E., Lu, V., Zhang, J., Chabra, S., Abishek, J., Wong, E., Zeng, X., & Guo, B. (2023). Using Pre-Clinical Studies to Explore the Potential Clinical Uses of Exosomes Secreted from Induced Pluripotent Stem Cell-Derived Mesenchymal Stem cells. *Tissue Engineering and Regenerative Medicine*. https://dx.doi.org/10.1007/s13770-023-00557-6

25. Zhou, Y., Zhao, B., Zhang, X.-L., Lu, Y.-J., Cheng, J., Fu, Y., Zhang, N., Li, P., Zhang, J., & Zhang, J. (2021). Toxicological Evaluation of Human Adipose-derived Mesenchymal Stem Cells (hADSCs) and hADSCs-derived Exosomes. *Research Square*. https://dx.doi.org/10.21203/rs.3.rs-626854 /v1

26. Mai, Z., Chen, H., Ye, Y., Hu, Z., Sun, W., Cui, L., & Zhao, X. (2021). Translational and Clinical Applications of Dental Stem Cell-Derived Exosomes. *Frontiers in Genetics*. https://dx.doi.org/10.3389/fgene.2021.750990

27. Hajinejad, M., & Sahab-Negah, S. (2021). Neuroinflammation: The next target of exosomal microRNAs derived from mesenchymal stem cells in the context of neurological disorders. *Journal of Cellular Physiology, 236*(8), 6241–6250. https://doi.org/10.1002/jcp.30495

28. Andrzejewska, A., Dąbrowska, S., Lukomska, B., & Janowski, M. (2021). Mesenchymal Stem Cells for Neurological Disorders. *Advanced Science, 8*(7), 2002944. https://doi.org/10.1002/advs.202002944

29. Ghafouri-Fard, S., Niazi, V., Hussen, B. M., Omrani, M., Taheri, M., & Basiri, A. (2021). The Emerging Role of Exosomes in the Treatment of Human Disorders With a Special Focus on Mesenchymal Stem Cells-Derived Exosomes. *Frontiers in Cell and Developmental Biology, 9*, 653296. https://doi.org/10.3389/fcell.2021.653296

30. Malekpour, K., Hazrati, A., Zahar, M., Markov, A., Zekiy, A., Gholizadeh Navashenaq, J., Roshangar, L., & Ahmadi, M. (2021). The Potential Use of Mesenchymal Stem Cells and Their Derived Exosomes for Orthopedic Diseases Treatment. *Stem Cell Reviews and Reports, 17*(3), 1022–1036. https://doi.org/10.1007/s12015-021-10185-z

31. Zheng, Q., Zhang, S., Guo, W., & Li, X.-K. (2021). The unique immunomodulatory properties of MSC-derived ex-

osomes in organ transplantation. *Frontiers in Immunology,* *12,* 659621. https://doi.org/10.3389/fimmu.2021.659621

32. Bulut, Ö., & Gürsel, İ. (2020). Mesenchymal stem cell derived extracellular vesicles: promising immunomodulators against autoimmune, autoinflammatory disorders and SARS-CoV-2 infection. *Turkish Journal of Biology, 44*(3), 273-282. https://doi.org/10.3906/biy-2002-79

33. Shen, Z., Huang, W., Liu, J., Tian, J., Wang, S., & Rui, K. (2021). Effects of mesenchymal stem cell-derived exosomes on autoimmune diseases. *Frontiers in Immunology,* *12,* 749192. https://doi.org/10.3389/fimmu.2021.749192

34. Zhu, X.-j., Ma, D., Yang, B., An, Q., Zhao, J., Gao, X., & Zhang, L. (2023). Research progress of engineered mesenchymal stem cells and their derived exosomes and their application in autoimmune/inflammatory diseases. *Stem Cell Research & Therapy, 14*(1), 95. https://doi.org/10.1186/s1 3287-023-03295-7

35. Théry, C., Amigorena, S., Raposo, G., & Clayton, A. (2006). Isolation and characterization of exosomes from cell culture supernatants and biological fluids. *Current Protocols in Cell Biology, Chapter 3,* Unit 3.22. https://doi.org/10.1002/04 71143030.cb0322s30

36. Jeppesen, D. K., et al. (2014). Comparative analysis of discrete exosome fractions obtained by differential centrifugation. *Journal of Extracellular Vesicles, 3.* https://doi.org/10 .3402/jev.v3.25011

37. Boing, A. N., et al. (2014). Single-step isolation of extracellular vesicles by size-exclusion chromatography. *Journal of Extracellular Vesicles, 3.* https://doi.org/10.3402/jev.v3.23430

38. Sidhom, K., Obi, P. O., & Saleem, A. (2020). A Review of Exosomal Isolation Methods: Is Size Exclusion Chromatography the Best Option? *International Journal of Molecular Sciences, 21*(18), 6466. https://doi.org/10.3390/ijms21186 466

39. Rider, M. A., Hurwitz, S. N., & Meckes, D. G. (2016). ExtraPEG: A Polyethylene Glycol-Based Method for Enrichment of Extracellular Vesicles. *Scientific Reports, 6,* 23978. https://doi.org/10.1038/srep23978

40. Helwa, I., et al. (2017). A Comparative Study of Different Methods for Endothelial Cell-Derived Extracellular Vesicles Isolation and Purification. *Journal of Circulating Biomarkers, 6.* https://doi.org/10.5772/67537

41. Taylor, D. D., & Shah, S. (2015). Methods of isolating extracellular vesicles impact down-stream analyses of their cargoes. *Methods, 87,* 3-10. https://doi.org/10.1016/j.ymeth. 2015.02.019

42. Alvarez, M. L., Khosroheidari, M., Kanchi Ravi, R., & DiStefano, J. K. (2012). Comparison of protein, microRNA, and mRNA yields using different methods of urinary exosome isolation for the discovery of kidney disease biomarkers. *Kidney International, 82*(9), 1024-1032. https://doi.o rg/10.1038/ki.2012.256

43. Chen, C., Skog, J., Hsu, C. H., Lessard, R. T., Balaj, L., Wurdinger, T., ... & Breakefield, X. O. (2010). Microfluidic isolation and transcriptome analysis of serum microvesicles. *Lab on a Chip, 10*(4), 505-511. https://doi.org/10.1039/b916199f

44. Shao, H., et al. (2018). New Technologies for Analysis of Extracellular Vesicles. *Chemical Reviews, 118*(4), 1917-1950. https://doi.org/10.1021/acs.chemrev.7b00534

45. Sung, S., Seo, M.-S., Kang, K., Choi, J.-H., Lee, S., Lim, J.-H., Yang, S., Kim, S.-K., & Lee, G. (2021). Isolation and Characterization of Extracellular Vesicle from Mesenchymal Stem Cells of the Epidural Fat of the Spine. *Asian Spine Journal.* https://dx.doi.org/10.31616/asj.2021.0129

46. Liu, X., Ren, F., Li, S., Zhang, N., Pu, J. J., Zhang, H., Xu, Z., Tan, Y., Chen, X., Chang, J., & Wang, H. (2023). Acute myeloid leukemia cells and MSC-derived exosomes inhibiting transformation in myelodysplastic syndrome. *Stem Cell Research & Therapy.* https://dx.doi.org/10.1007/s12672-023-00714-2

47. Liu, P., Zhang, Q., Mi, J., Wang, S., Xu, Q., Zhuang, D., Chen, W., Liu, C., Zhang, L., Guo, J., & Wu, X. (2022). Exosomes derived from stem cells of human deciduous exfoliated teeth inhibit angiogenesis in vivo and in vitro via the transfer of miR-100-5p and miR-1246. *Stem Cell Research & Therapy.* https://dx.doi.org/10.1186/s13287-022-02764-9

48. Wang, A., Liu, J., Yu, S., Liu, X.-M., Zhuang, X., Liu, Y., &

Chen, X. (2021). Exosomes Derived from Stem Cells from Apical Papilla Ameliorate Sjogren's Syndrome by Suppressing Th17 Cell Differentiation. *Research Square*. https://dx .doi.org/10.21203/rs.3.rs-864685/v1

49. Andriolo, G., Provasi, E., Lo Cicero, V., Brambilla, A., Soncin, S., Torre, T., Milano, G., Biemmi, V., Vassalli, G., Turchetto, L., Barile, L., & Radrizzani, M. (2018). Exosomes From Human Cardiac Progenitor Cells for Therapeutic Applications: Development of a GMP-Grade Manufacturing Method. *Frontiers in Physiology*. https://dx.doi. org/10.3389/fphys.2018.01169

50. Franquesa, M., Hoogduijn, M., Ripoll, É., Luk, F., Salih, M., Betjes, M., Torras, J., Baan, C., Grinyó, J., & Merino, A. (2014). Update on Controls for Isolation and Quantification Methodology of Extracellular Vesicles Derived from Adipose Tissue Mesenchymal Stem Cells. *Frontiers in Immunology*. https://dx.doi.org/10.3389/fimmu.2014.00525

51. Yi, T., Kim, S.-n., Lee, H., Kim, J., Cho, Y., Shin, D.-H., Tak, S.-J., Moon, S., Kang, J., Ji, I.-M., Lim, H.-J., Lee, D.-S., Jeon, M.-S., & Song, S. U. (2015). Manufacture of Clinical-Grade Human Clonal Mesenchymal Stem Cell Products from Single Colony Forming Unit-Derived Colonies Based on the Subfractionation Culturing Method. *Tissue Engineering Part C: Methods*. https://dx.doi.org/10.1089/ten. TEC.2015.0017

52. Ducret, M., Fabre, H., Degoult, O., Atzeni, G., McGuckin, C., Forraz, N., Mallein-Gerrin, F., Perrier-Groult, E., & Far-

gues, J. (2016). A standardized procedure to obtain mesenchymal stem/stromal cells from minimally manipulated dental pulp and Wharton's jelly samples. *PubMed*. https://pubmed.ncbi.nlm.nih.gov/27352427

53. Schirmaier, C., Jossen, V., Kaiser, S., Jüngerkes, F., Brill, S., Safavinab, A., Siehoff, A., Bos, C., Eibl, D., & Eibl, R. (2014). Scale-up of adipose tissue-derived mesenchymal stem cell production in stirred single-use bioreactors under low-serum conditions. *Engineering in Life Sciences*, 14(3), 292-303. https://dx.doi.org/10.1002/elsc.201300134

54. □□□□□□□□□□□, □□ □□, Teruya, K., & Katakura, Y. (Eds.). (2001). Animal cell technology: Basic & applied aspects: Proceedings of the Thirteenth Annual Meeting of the Japanese Association for Animal Cell Technology (JAACT), Fukuoka-Karatsu, November 16-21, 2000. [No DOI available]

55. Andriolo, G., Provasi, E., Lo Cicero, V., Brambilla, A., Soncin, S., Torre, T., Milano, G., Biemmi, V., Vassalli, G., Turchetto, L., Barile, L., & Radrizzani, M. (2018). Exosomes From Human Cardiac Progenitor Cells for Therapeutic Applications: Development of a GMP-Grade Manufacturing Method. *Frontiers in Physiology, 9*, 1169. https://doi.org/10.3389/fphys.2018.01169

56. Chen, Y.-S., Lin, E., Chiou, T., & Harn, H. (2019). Exosomes in clinical trial and their production in compliance with good manufacturing practice. *Tzu Chi Medical Journal, 31*(4), 255–259. https://doi.org/10.4103/tcmj.tcmj_1

82_19

57. Sensébé, L., Gadelorge, M., & Fleury-Cappellesso, S. (2013). Production of mesenchymal stromal/stem cells according to good manufacturing practices: A review. *Stem Cell Research & Therapy, 4*(3), 66. https://doi.org/10.1186/scrt217

58. Paulitti, A., Barchiesi, A., Moretti, M., & Cattaruzzi, G. (n.d.). Automated Expansion of Adipose-Derived Mesenchymal Stem Cells (AD-MSCs) with NANT 001 System. ↩

59. Perico, N., Casiraghi, F., & Remuzzi, G. (2017). Clinical Transplantation of Mesenchymal Stromal Cell Therapies in Nephrology. Journal of the American Society of Nephrology, 29, ccc–ccc. https://doi.org/10.1681/ASN.2017070781 ↩

60. Aiastui, A. (2015). Should Cell Culture Platforms Move towards EV Therapy Requirements? Frontiers in Immunology, 6, 8. https://doi.org/10.3389/fimmu.2015.00008 ↩

61. Tonn, T., & Barz, D. (2008). MSC – a Multipotent Stromal Cell in Search of Clinical Application. Transfusion Medicine and Hemotherapy, 35(4), 269-279. https://doi.org/10.1159/000147276 ↩

GLOSSARY

1. **Exosomes**: Small, membrane-bound vesicles released by cells, including mesenchymal stem cells (MSCs), involved in cell-to-cell communication, measuring about 30-150 nanometers in diameter.

2. **Mesenchymal Stem Cells (MSCs)**: A type of adult stem cell found in various tissues capable of differentiating into multiple cell types and possessing immunomodulatory properties.

3. **Endosomal Compartment**: A membrane-bound compartment within the cell where exosomes are formed through inward budding.

4. **Multivesicular Bodies (MVBs)**: Cellular structures containing intraluminal vesicles that become exosomes when they fuse with the plasma membrane.

5. **Intraluminal Vesicles**: Small vesicles contained within MVBs that are released as exosomes.

6. **Plasma Membrane**: The outer membrane of a cell that can fuse with MVBs to release exosomes.

7. **Bioactive Molecules**: Active biological compounds, including proteins, lipids, nucleic acids, and signaling molecules, that can be carried by exosomes.

8. **Nucleic Acids**: Biological molecules, such as DNA, mRNA, and microRNAs, that can be part of exosome cargo.

9. **Immunomodulatory**: Having the ability to modify or regulate one or more immune functions.

10. **Regenerative Medicine**: A field of medicine that develops methods to regrow, repair, or replace damaged or diseased cells, organs, or tissues.

11. **Angiogenesis**: The formation of new blood vessels, a process that can be promoted by MSC-derived exosomes.

12. **Neuroplasticity**: The ability of the nervous system to change its activity in response to intrinsic or extrinsic stimuli by reorganizing its structure, functions, or connections.

13. **Myocardial Infarction**: Commonly known as a heart attack, a medical condition that occurs when blood flow to the heart is blocked.

14. **Graft-versus-Host Disease**: A complication following a transplant where the donor's immune cells attack the recip-

ient's body tissues.

15. **Good Manufacturing Practices (GMP)**: Guidelines that provide the requirements that a manufacturer must meet to ensure products are of high quality and do not pose any risk to the consumer.

16. **Nanoparticle Tracking Analysis**: A technique used to characterize the size distribution and concentration of nanoparticles in a liquid suspension, such as exosomes.

17. **Transmission Electron Microscopy (TEM)**: A microscopy technique in which a beam of electrons is transmitted through a specimen to form an image.

18. **Proteomic Analysis**: The large-scale study of proteins, particularly their structures and functions.

19. **Clinical Trials**: Research studies performed in people that are aimed at evaluating a medical, surgical, or behavioral intervention.

20. **Preclinical Studies**: Research studies conducted to test a drug, procedure, or other medical treatment in animals before trials are carried out in humans.

21. **Investigational New Drug (IND) Application**: A request for authorization from the FDA to administer an investigational drug or biological product to humans.

22. **Ultracentrifugation**: A process where a centrifuge is used to spin samples at very high speeds to separate components based on density.

23. **Density Gradient Centrifugation**: A technique used to separate particles based on their size, shape, and density by spinning them in a fluid that has a gradient of density.

24. **Size Exclusion Chromatography**: A method to separate molecules based on their size using a gel-like column.

25. **Immunoaffinity Capture**: A technique to isolate molecules from a mixture based on the specific and reversible interaction between an antigen and its antibody.

26. **Dynamic Light Scattering (DLS)**: A technique used to determine the size distribution of small particles in suspension or polymers in solution.

27. **Western Blotting**: A laboratory method used to detect specific protein molecules from a mixture of proteins.

28. **Flow Cytometry**: A technology that is used to analyze the physical and chemical characteristics of particles in a fluid as it passes through at least one laser.

29. **Mass Spectrometry**: An analytical technique that measures the mass-to-charge ratio of ions to identify and quantify molecules in a sample.

30. **Exosomal Cargo**: The diverse array of molecules encapsulated within exosomes, responsible for their therapeutic potential and intercellular communication functions.

31. **Mechanisms of Action**: The specific biochemical interaction through which a drug substance produces its pharmacological effect.

32. **Sorting Mechanisms**: Biological processes that determine which molecules are packaged into exosomes, affecting the composition of the cargo.

33. **Surface Markers**: Molecules present on the surface of exosomes used for their identification and isolation, such as CD9, CD63, and CD81.

34. **Cell Adhesion**: The process by which cells interact and attach to a surface, substrate, or another cell, mediated by interactions at the cell surface.

35. **Signaling Pathways**: A group of molecules in a cell that work together to control one or more cell functions, such as cell division or cell death.

36. **Cholesterol**: A type of lipid that is an essential structural component of all animal cell membranes and is involved in the formation and maintenance of membrane fluidity.

37. **Sphingomyelin**: A type of sphingolipid found in animal cell membranes, especially in the membranous myelin sheath that surrounds some nerve cell axons.

38. **Phospholipids**: A class of lipids that are a major component of all cell membranes as they can form lipid bilayers.

39. **MicroRNA (miRNA)**: Small non-coding RNA molecules found in plants, animals, and some viruses, which function in RNA silencing and post-transcriptional regulation of gene expression.

40. **Non-coding RNAs**: RNA molecules that are not translated

into proteins but can have roles in gene expression and regulation.

41. **Growth Factors**: A group of proteins that stimulate the growth of specific tissues.

42. **Cytokines**: Small proteins that are important in cell signaling. They are released by cells and affect the behavior of other cells.

43. **Chemokines**: A family of small cytokines, or signaling proteins secreted by cells. Their name is derived from their ability to induce directed chemotaxis in nearby responsive cells.

44. **Gene Expression**: The process by which information from a gene is used in the synthesis of a functional gene product, such as proteins, which can affect the cell's structure and function.

45. **Osteoblasts**: Cells with a single nucleus that synthesize bone.

46. **Cardiomyocytes**: The cells that make up the cardiac muscle (heart muscle).

47. **Regulatory Immune Cells**: A variety of immune cell types that maintain immune tolerance and prevent autoimmune disease by suppressing excessive immune responses.

48. **Targeted Drug Delivery**: A method of delivering medication to a patient in a manner that increases the concentration of the medication in some parts of the body relative to others.

49. **Gene Therapy**: A technique that uses genes to treat or prevent disease.

50. **Ischemia**: A restriction in blood supply to tissues, causing a shortage of oxygen that is needed for cellular metabolism.

51. **Cargo Heterogeneity**: The variation in the composition of molecules within exosomal cargo, which can affect the therapeutic efficacy and consistency of exosomal preparations.

52. **Cargo Loading and Modification**: The process of incorporating specific therapeutic molecules into exosomes, which can be challenging and may introduce additional complexities.

53. **Scalability**: The capability of a process to be used or produced in a range of capabilities without losing functionality.

54. **Immunogenicity**: The ability of a particular substance, such as an antigen or epitope, to provoke an immune response in the body of a human or animal.

55. **Off-target Effects**: Unintended actions of a drug that occur in addition to the desired therapeutic effect.

56. **Clinical Translation**: The process of applying discoveries generated during research in the laboratory, and in preclinical studies, to the development of trials and studies in humans.

57. **Standardization**: The process of implementing and developing technical standards to maximize compatibility, repeatability, safety, and quality.

58. **Exosomes**

59. : Small extracellular vesicles secreted by cells, including mesenchymal stem cells (MSCs), that facilitate intercellular communication by transferring molecules such as proteins, lipids, and nucleic acids to recipient cells.

60. **Mesenchymal Stem Cells (MSCs)**: A type of adult stem cell found in various tissues capable of differentiating into multiple cell types and possessing immunomodulatory and regenerative properties.

61. **Tissue Repair and Regeneration**: The process by which cells repair and regenerate tissue, often facilitated by MSC-derived exosomes in various organs.

62. **Wound Healing**: The body's natural process of repairing tissue damage, in which MSC-derived exosomes can accelerate the process by promoting cell migration and reducing inflammation.

63. **Bone and Cartilage Regeneration**: The process of bone and cartilage formation, which can be stimulated by MSC-derived exosomes that encourage the differentiation of osteoblasts and chondrocytes.

64. **Neurological Disorders**: Medical conditions that affect the brain, spinal cord, and nerves, where MSC-derived exosomes may have therapeutic applications by promoting neuronal survival and modulating the inflammatory response.

65. **Neuroprotection**: The preservation of neuronal structure and function, which can be supported by MSC-derived ex-

osomes in various neurological disorders.

66. **Neuroregeneration**: The regrowth or repair of nervous tissues, cells, or cell products, which MSC-derived exosomes can facilitate.

67. **Neuroplasticity**: The ability of the nervous system to change its activity in response to intrinsic or extrinsic stimuli by reorganizing its structure, functions, or connections.

68. **Immunomodulatory**: The capability to modify or regulate one or more immune functions, a property exhibited by MSC-derived exosomes.

69. **Cardiovascular Diseases (CVDs)**: A class of diseases that involve the heart or blood vessels where MSC-derived exosomes may help in tissue repair and reducing inflammation.

70. **Myocardial Infarction**: Commonly known as a heart attack, it occurs when blood flow decreases or stops to a part of the heart, causing damage to the heart muscle.

71. **Ischemic Heart Disease**: A disease characterized by reduced blood supply to the heart.

72. **Heart Failure**: A chronic condition in which the heart doesn't pump blood as well as it should.

73. **Angiogenesis**: The development of new blood vessels, a process that can be enhanced by MSC-derived exosomes in cardiac tissue repair.

74. **Oxidative Stress**: An imbalance between free radicals and

antioxidants in the body, which can lead to cell and tissue damage.

75. **Immune Modulation**: The process of regulating the immune system to achieve a therapeutic effect, for which MSC-derived exosomes can be utilized.

76. **Autoimmune Diseases**: Diseases in which the body's immune system attacks healthy cells, potentially treatable by the immunomodulatory effects of MSC-derived exosomes.

77. **Graft-versus-Host Disease**: A complication following a bone marrow or stem cell transplant in which the donated bone marrow or peripheral blood stem cells view the recipient's body as foreign and attack the body.

78. **Inflammatory Bowel Disease (IBD)**: Chronic inflammation of the digestive tract, including Crohn's disease and ulcerative colitis.

79. **Immunosuppressive Drugs**: Medications that lower the body's immune response, potentially replaceable by MSC-derived exosomes to promote immune tolerance in organ transplantation.

80. **Good Manufacturing Practices (GMP)**: A system for ensuring that products are consistently produced and controlled according to quality standards. It is designed to minimize the risks involved in any pharmaceutical production that cannot be eliminated through testing the final product.

81. **Regulatory Authorities**: Governmental bodies that regulate various sectors, with the FDA (Food and Drug Adminis-

tration) in the United States and the EMA (European Medicines Agency) in Europe being examples for pharmaceuticals. They are responsible for the evaluation and monitoring of safety, efficacy, and quality of drugs.

82. **Facility Design and Control:** Refers to the specifications and operations of a manufacturing facility to ensure it meets the necessary standards for the production process, particularly in maintaining a clean and controlled environment.

83. **Equipment and Instrumentation:** Tools and devices used in the manufacturing process that must be regularly checked and maintained to ensure they are functioning correctly and providing accurate results.

84. **Raw Material Control:** The management and oversight of materials used in the production process, ensuring they meet certain standards and are handled in a way that maintains their quality.

85. **Standard Operating Procedures (SOPs):** Detailed, written instructions to achieve uniformity of the performance of a specific function.

86. **Personnel Training and Qualification:** Ensuring that individuals involved in the manufacturing process are properly educated and trained to perform their duties effectively and in compliance with GMP standards.

87. **Quality Control and Testing:** The part of GMP concerned with sampling, specifications, and testing, and with the organization, documentation, and release procedures

which ensure that the necessary and relevant tests are carried out.

88. **Documentation and Record-Keeping:** Maintaining records throughout the manufacturing process to ensure traceability and compliance with GMP standards.

89. **Risk Assessment:** The process of identifying potential risks in the manufacturing process and determining their severity and the likelihood of their occurrence.

90. **Donor Screening and Testing:** The process of evaluating potential donors of biological materials (like MSCs) for risk factors and infectious diseases to ensure the safety of the biological products derived from them.

91. **Exosome Isolation and Purification:** The methods used to separate exosomes from other cellular components, which are crucial for ensuring the purity and quality of the exosomes for therapeutic use.

92. **Characterization of Exosomes:** The analysis of exosomes to determine their size, structure, surface markers, and content, which is essential for ensuring their consistency and functionality.

93. **Investigational New Drug (IND) Application:** A request for authorization from the FDA to administer an investigational drug to humans, which is necessary before beginning clinical trials.

94. **Clinical Trials:** Research studies performed in people that are aimed at evaluating a medical, surgical, or behavioral in-

tervention.

95. **Regulatory Approval:** The process by which a regulatory agency (such as the FDA or EMA) evaluates the safety and efficacy of a pharmaceutical product to determine whether it can be sold and marketed.

96. **Post-Marketing Surveillance:** The monitoring of pharmaceutical products after they have been released on the market to ensure continued safety and efficacy.

97. **International Harmonization and Standardization:** Efforts by international bodies to align regulatory requirements, standards, and procedures across different regions to facilitate global development and approval of pharmaceutical products.

98. **International Society for Extracellular Vesicles (ISEV):** A professional organization that aims to advance the study of extracellular vesicles, including exosomes, on a global scale.

99. **International Organization for Standardization (ISO):** An independent, non-governmental international organization that develops standards to ensure the quality, safety, and efficiency of products, services, and systems.

100. **Good Manufacturing Practices (GMP):** A system for ensuring that products are consistently produced and controlled according to quality standards. It is designed to minimize the risks involved in any pharmaceutical production that cannot be eliminated through testing the final product.

101. **Regulatory Authorities:** Governmental bodies that regu-

late various sectors, with the FDA (Food and Drug Administration) in the United States and the EMA (European Medicines Agency) in Europe being examples for pharmaceuticals. They are responsible for the evaluation and monitoring of safety, efficacy, and quality of drugs.

102. **Facility Design and Control:** Refers to the specifications and operations of a manufacturing facility to ensure it meets the necessary standards for the production process, particularly in maintaining a clean and controlled environment.

103. **Equipment and Instrumentation:** Tools and devices used in the manufacturing process that must be regularly checked and maintained to ensure they are functioning correctly and providing accurate results.

104. **Raw Material Control:** The management and oversight of materials used in the production process, ensuring they meet certain standards and are handled in a way that maintains their quality.

105. **Standard Operating Procedures (SOPs):** Detailed, written instructions to achieve uniformity of the performance of a specific function.

106. **Personnel Training and Qualification:** Ensuring that individuals involved in the manufacturing process are properly educated and trained to perform their duties effectively and in compliance with GMP standards.

107. **Quality Control and Testing:** The part of GMP concerned with sampling, specifications, and testing, and with

the organization, documentation, and release procedures which ensure that the necessary and relevant tests are carried out.

108. **Documentation and Record-Keeping:** Maintaining records throughout the manufacturing process to ensure traceability and compliance with GMP standards.

109. **Risk Assessment:** The process of identifying potential risks in the manufacturing process and determining their severity and the likelihood of their occurrence.

110. **Donor Screening and Testing:** The process of evaluating potential donors of biological materials (like MSCs) for risk factors and infectious diseases to ensure the safety of the biological products derived from them.

111. **Exosome Isolation and Purification:** The methods used to separate exosomes from other cellular components, which are crucial for ensuring the purity and quality of the exosomes for therapeutic use.

112. **Characterization of Exosomes:** The analysis of exosomes to determine their size, structure, surface markers, and content, which is essential for ensuring their consistency and functionality.

113. **Investigational New Drug (IND) Application:** A request for authorization from the FDA to administer an investigational drug to humans, which is necessary before beginning clinical trials.

114. **Clinical Trials:** Research studies performed in people that

are aimed at evaluating a medical, surgical, or behavioral intervention.

115. **Regulatory Approval:** The process by which a regulatory agency (such as the FDA or EMA) evaluates the safety and efficacy of a pharmaceutical product to determine whether it can be sold and marketed.

116. **Post-Marketing Surveillance:** The monitoring of pharmaceutical products after they have been released on the market to ensure continued safety and efficacy.

117. **International Harmonization and Standardization:** Efforts by international bodies to align regulatory requirements, standards, and procedures across different regions to facilitate global development and approval of pharmaceutical products.

118. **International Society for Extracellular Vesicles (ISEV):** A professional organization that aims to advance the study of extracellular vesicles, including exosomes, on a global scale.

119. **International Organization for Standardization (ISO):** An independent, non-governmental international organization that develops standards to ensure the quality, safety, and efficiency of products, services, and systems.

www.ingramcontent.com/pod-product-compliance
Lightning Source LLC
Chambersburg PA
CBHW071209290526
45796CB00008B/190